Franz Keibel

Normentafeln zur Entwicklungsgeschichte der Wirbeltiere

Band 11

Franz Keibel

Normentafeln zur Entwicklungsgeschichte der Wirbeltiere
Band 11

ISBN/EAN: 9783743322929

Hergestellt in Europa, USA, Kanada, Australien, Japan

Cover: Foto ©berggeist007 / pixelio.de

Manufactured and distributed by brebook publishing software
(www.brebook.com)

Franz Keibel

Normentafeln zur Entwicklungsgeschichte der Wirbeltiere

NORMENTAFELN

ZUR

ENTWICKLUNGSGESCHICHTE DER WIRBELTIERE.

IN VERBINDUNG MIT

Dr. Bles-Glasgow, Dr. Boeke-Hebber, Holland, Prof. Dr. Brachet-Brüssel, Prof. Dr. B. Dean-Columbia University, New York, U. S. A., Dr. H. Fuchs-Strassburg, Dr. Glaesner-Strassburg, Prof. Dr. O. Grosser-Prag, Prof. Dr. B. Henneberg-Giessen, Prof. Dr. Hubrecht-Utrecht, Prof. J. Graham Kerr-Glasgow, Prof. Dr. Kopsch-Berlin, Dr. Thilo Krumbach-Breslau, Prof. Dr. Ludosch-Jena, Prof. Dr. P. Martin-Giessen, Dr. Nierstrasz-Utrecht, Prof. Dr. C. S. Minot-Boston, U. S. A., Prof. Dr. Nicolas-Paris, Prof. Dr. Peter-Greifswald, Prof. Reighard-Ann Arbor, U. S. A., Dr. Sakurai-Fukuoka, Japan, Dr. Scammon-Harvard Medical School, Boston, U. S. A., Prof. Dr. Semon-Prinz-Ludwigshöhe bei München, Prof. Dr. Sobotta-Würzburg, Prof. Dr. Soulié-Toulouse, Prof. Dr. Tandler-Wien, Dr. Taylor-Philadelphia, U. S. A., Prof. Dr. Tourneux, Toulouse, Dr. Voelker-Prag, Prof. Whitman-Chicago, U. S. A.

HERAUSGEGEBEN VON

PROF. DR. F. KEIBEL, LL. D. (HARVARD),

FREIBURG I. BR.

ELFTES HEFT.

NORMAL PLATES OF THE DEVELOPMENT OF NECTURUS MACULOSUS.

BY

ALBERT C. EYCLESHYMER and JAMES M. WILSON.

ST. LOUIS UNIVERSITY, ST. LOUIS MO., U. S. A.

WITH 3 PLATES.

JENA,

VERLAG VON GUSTAV FISCHER.

1910.

Table of Contents.

Description of Illustrations.

The series of eggs, embryos and larvae of *Necturus*, from which the following descriptions and the appended illustrations were made, were collected May 15th, 1903 and kept at a water temperature of 17° 18° C. The illustrations are copied from the original water colored pictures which were made by Mr. Leonard H. Wilder, under the direction of the senior author. It should be emphasized that the ages, measurements and illustrations are all made from the living objects.

Fig. 1. (× 10.)

Side view of egg 1 day 4 hrs. after deposition. The first cleavage groove has reached the lower pole of the egg. Second grooves extend to level of the equator of the egg.

Fig. 2. (× 10.)

Side view of egg 1 day 8 hrs. after deposition. The second cleavage grooves have reached the equator. The grooves of the third cleavage pass in meridional planes, but have not yet reached the equator.

Fig. 3. (× 10.)

Side view of egg 1 day 12 hrs. old. Five cleavage grooves have reached lower pole, dividing lower hemisphere into six segments.

Fig. 4. (× 10.)

Side view of egg 1 day 16 hrs. old. The greater number of cleavage grooves pass in meridional planes, many are latitudinal and some nearly radial. The upper surface of the egg shows sixteen segments, the lower nine.

Fig. 5. (× 10.)

Side view of egg 1 day 20 hrs. old. The upper surface of the egg shows some fifty segments, the lower nine.

Fig. 6. (× 10.)

Side view of egg 2 days 2 hrs. old. The upper surface of the egg shows more than one hundred segments, the lower twelve.

Fig. 7. (× 10.)

Side view of egg 2 days 7 hrs. old. The upper surface of egg shows about two hundred cells. The lower portion is in about same stage as described in Fig. 6.

Fig. 8. (× 10.)

Side view of egg 2 days 12 hrs. old. The upper surface of egg shows some five hundred cells, the lower about forty.

Fig. 9. (× 10.)

Top view of egg 4 days 4 hrs. old. Segmentation cavity shows through thin translucent roof. Blastopore not present.

Fig. 10. (× 10.)

Bottom view of egg 6 days 16 hrs. old. Crescentic blastopore. Line of invagination sharply separates large yolk cells from small cells of blastodisc.

Fig. 11. (× 10.)

Dorso-lateral view of egg 10 days 10 hrs. old. Large circular blastopore; faint indication of embryonic anlage.

Fig. 12. (× 10.)

Side view of egg 10 days 16 hrs. old. Large circular blastopore. Anlage of mesial portion of embryo above dorsal lip of blastopore. Segmentation cavity faintly outlined.

Fig. 13. (× 10.)

Top view of egg 13 days 3 hrs. old. Small circular blastopore. Embryonic anlage triangular in outline; lateral boundaries indistinct. First appearance of neural groove. Roof of segmentation cavity thinner, making its boundaries distinct.

Fig. 14. (× 10.)

Top view of egg 14 days 4 hrs. old. Blastopore smaller, lateral margins of anterior portion of embryo bounded by short broad ridges which are the beginnings of the lateral portions of the neural fold. At anterior margin of embryo there is a transverse crescentic ridge which is beginning of transverse portion of neural fold. Neural groove deep but does not extend either to transverse portion of neural fold or to blastopore. Segmentation cavity crescentic.

Fig. 15. (× 10.)

Top view of egg 14 days 19 hrs. old. Blastopore much reduced, circular. The yolk plug is not visible in this egg. Lateral and transverse portions of neural fold united to form continuous fold around anterior portion of embryo. Lateral boundaries of posterior portion of embryo not defined. Neural groove not as long, nor as distinct as in preceding stage. Dark crescentic area in front of embryo is segmentation cavity.

Fig. 16. (× 10.)

Top view of egg 15 days 15 hrs. old. Blastopore small, circular; yolk plug visible. Neural fold prominent, its free ends extend nearly to blastopore. Neural groove deep and narrow at anterior end, broad and shallow at posterior end, fades out just in front of blastopore. A part of the segmentation cavity is still apparent in front of the embryo.

Fig. 17. (× 10.)

Top view of egg 16 days 6 hrs. old. Blastopore reduced to a very minute circular aperture. Neural plate narrower than in preceding stage. Neural fold prominent, its free ends coalescing at blastopore.

Neural groove extends to transverse portion of fold but does not reach blastopore. Segmentation cavity no longer visible in surface views.

Fig. 18. (X 10.)

Top view of embryo 17 days 2 hrs. old. Blastopore an elongated narrow aperture between ends of neural fold. Neural plate narrower than in preceding stage. The constricted portion represents in a general way the division between head and trunk. Neural fold most prominent in head region.

Fig. 19. (X 10.)

Top view of egg 17 days 17 hrs. old. Blastopore no longer visible. Neural plate narrowest posteriorly; broad in head region, showing boundary zone between head and trunk. Lateral portions of fold coalesced at posterior end of embryo. At anterior end of embryo a deep groove partially separates the two halves of the neural fold.

Fig. 20. (X 10.)

Top view of egg 18 days 13 hrs. old. Lateral portions of neural fold almost united except in head region where they are still widely separated. In the antero-lateral portions of the fold are slight evaginations which are the beginnings of the optic vesicles.

Fig. 21. (X 10.)

Top view of egg 18 days 15 hrs. old, 3 or 4 pairs of myotomes. Lateral portions of neural fold widely separated in head region, more closely approximated in anterior trunk region, coalesced in tail.

Fig. 22. (X 10.)

Dorso-lateral view of embryo 20 days 10 hrs. old, length 6 mm, 6 pairs of myotomes. Outline of body conforms to curvature of egg. Head end of embryo shows three longitudinal ridges; middle ridge lies slightly above level of lateral ridges. The middle one is common anlage of fore, mid and hind brain. The lateral ones are the common anlage of the optic vesicles and branchial arches. Anus formed.

Fig. 23. (X 10.)

Side view of embryo 21 days 2 hrs. old, length 7 mm, 10—12 pairs of myotomes. General outline of body conforms to curvature of egg. Head slightly raised above surface of yolk. Slight enlargement at end of tail. A distinct enlargement of anterior end of head shows optic vesicles; just posterior to this enlargement is the anlage of the branchial arches. Anus shows just below tip of tail.

Fig. 24. (X 10.)

Dorso-lateral view of embryo 22 days 17 hrs. old, length 8 mm, 16—18 pairs of myotomes. Embryo much curved laterally. Anterior half of head free from yolk. Caudal enlargement more prominent. Optic vesicles and mandibular arch well defined. The hyoid and first branchial arches are discernible; also the common anlage of the second and third branchial arches.

Fig. 25. (X 5.)

Side view of embryo 23 days 10 hrs. old, length 9 mm, 20—22 pairs of myotomes. General outline of the body straighter. Head free from yolk. Caudal enlargement becoming free. Optic vesicles and forebrain much larger. Mandibular, hyoid, first branchial, and common anlage of second and third branchial arches well defined. Otic vesicle visible above hyoid arch.

1*

Fig. 26. (X 5.)

Side view of embryo 24 days 22 hrs. old, length 10 mm, 23—24 pairs of myotomes. General outline of body of embryo straighter, less curved laterally. Head and caudal extremities free from yolk. Yolk becoming oval. Optic vesicles prominent. Ear better defined. Olfactory pits present. The mandibular, hyoid and first branchial arches are distinct. The second and third branchial arches are not yet differentiated, a slight process on the first branchial indicates the beginning of the gill bar. The anlage of the heart is visible just beneath the arches.

Fig. 27. (X 5.)

Side view of embryo 26 days old, length 11 mm, 26—27 myotomes. General outline of body straighter than in preceding stage. Head projects some 3 mm beyond margin of yolk; tail projects 1.2 mm, is thinner laterally but broader dorso-ventrally. Eye, ear, nasal pits and mouth well defined. Maxillary process discernible. Mandibular arches longer, but ventral ends widely separated. Second and third branchial arches formed. Gill bars present on three branchial arches. Anterior limb buds indicated; faint anlage of posterior limb buds. Yolk pear-shaped. Heart prominent. First surface capillaries present although not indicated in figure.

Fig. 28. (X 5.)

Side view of embryo 30 days 8 hrs. old, length 13 mm, 30—31 myotomes. The trunk of the embryo is nearly straight. At level of the posterior gill there is a pronounced neck bend and at the level of the posterior limbs a striking downward bend of the tail. The epiphysis shows in surface views. The lens is discernible. The ear is still visible. The external nasal openings are sharply defined. The boundaries of the mouth are better outlined owing to the approximation of the ventral ends of the mandibular arches. The hyoid arch is becoming obscured. The gill bars are prominent on the three branchial arches. The anterior limb buds project dorsally about .5 mm above the surface of the body. The posterior limb buds are but slight elevations. The yolk is pear-shaped with its dorsal surface much flattened. The auricular and ventricular portions of the heart are apparent. The surface of the yolk is covered by a dense network of capillaries which for the most part convey blood antero-ventrally to the abdominal vein. Considerable pigment is present in the trunk region although but little has reached the outer portion of the dermis.

Fig. 29. (X 5.)

Side view of embryo 36 days 16 hrs. old, length 16 mm, 36—38 myotomes. In general outline the embryo shows a number of striking changes. The neck bend is not so pronounced. The tail bend is scarcely noticeable. There is a striking increase in dorso-ventral width of tail. The cerebral hemispheres are well defined. The eye is now prominent and the lens better defined. The ear is no longer visible in surface views. The mouth is well defined. The ends of the mandibular arches are closely approximated but not united. The hyoid and branchial arches are more obscure. Anlagen of gill filaments present on gill bars. Anterior limbs project dorsally. Posterior limbs are short ridges extending in horizontal plane. The yolk is elongated and reduced in diameter both dorso-ventrally and laterally. Surface blood vessels as in preceding stage, excepting that they are now apparent in the gill bars. The chromatophores are most numerous in the anterior and dorsal portions of the head.

Fig. 30. (× 5.)

Side view of embryo 40 days 20 hrs. old, length 18 mm, 44–46 myotomes. The outline of the body shows a marked ventral curvature of the trunk, less pronounced neck bend, and further increase in the dorso-ventral width of the tail. The eye is very prominent owing to the pigment in the retina. Ear not visible externally. Nasal openings very small. The mandibular arches have coalesced. The boundaries of the other arches are no longer discernible. Gill filaments well developed. Anterior limbs about 1 mm long project dorso-posteriorly. The yolk is elongated oval. Abdominal vein and branchial blood vessels prominent. Pigment present in dorsal portion of head, also along dorsal and lateral portions of trunk and tail. The yolk is unpigmented excepting along dorsal margin.

Fig. 31. (× 5.)

Side view of larva 49 days old, length 21 mm. General outline of body decidedly different. Head bend obliterated, slight upward curve in trunk. Tail broader. Eye more deeply pigmented. Gill bars very long, extending to level of end of anterior limb. From three to five lateral filaments on each gill bar. Anterior limbs project postero-ventrally; three digits formed. Posterior limbs directed caudad; no trace of digits. Yolk much elongated. Network of capillaries denser. Large lateral arteries, at level of upper margin of yolk, very prominent. Well defined longitudinal bands of pigment.

Fig. 32. (× 5.)

Side view of larva 61 days old, length 25 mm. General outline of body shows less dorsal curvature of trunk. Tail much longer in proportion to length of trunk and much broader dorso-ventrally. Gill bars longer, each possessing six to eight lateral filaments. Anterior and posterior limbs directed postero-ventrally. Anterior 3 mm long, posterior 2 mm long. Each limb shows four digits. The distribution of pigment is essentially similar to that observed in the 21 mm larva, the bands however are more sharply defined. Chromatophores in the gill bars and limbs and beginning to extend over the dorsal surface of the yolk.

Fig. 33. (× 5.)

Side view of larva 70 days 4 hrs. old, length 28 mm. The general outline of the body is slenderer than at any time preceding. The rapid absorption of the yolk has brought its ventral surface nearly to the level of the ventral surfaces of the head and tail. The gill bars curve dorsally and possess from ten to twelve pairs of lateral filaments. The tail is somewhat constricted at the level of the posterior limbs. The limbs and digits are better developed and are now used in locomotion. Pigmentation is denser than in 25 mm larva, but same general arrangement of bands prevails.

Fig. 34. (× 5.)

Side view of larva 97 days old, length 34 mm. In general outline the larva begins to resemble the adult. The yolk is well absorbed. The tail is very broad and now used as a powerful caudal fin in swimming. The gill bars project dorsally and have a large number of filaments. The legs project far below the ventral surface of the body. In color the same general pattern prevails as in the 28 mm larva.

There are some minor changes, the light band is broader and better defined, and extensions of pigment over the yolk have been so uneven that a number of irregular oval areas are left unpigmented, causing a mottled appearance in this region.

Fig. 35. (× 5.)

Side view of larva 126 days old, length 39 mm. The young *Necturus* now conforms in outline to the adult. In color however it is decidedly different.

Introduction to Tables.

The material upon which the normal tables are based was collected in 1903. At the time the illustrations were made from the living material, several specimens of the same stages were fixed in various solutions. Among these 10 % formalin caused the least distortion. This formalin-fixed material proved most satisfactory for work not involving cytological study. The material was stained *in toto* with haematoxylin, imbedded in paraffin and counterstained on the slide with weak picric acid, orange G or eosin. Unless otherwise stated the above fixation and staining have been used. Each of the stages designated in the following tables was sectioned in transverse, horizontal and sagittal planes. Besides these series many others have been consulted in which the material was fixed and stained by other methods. The total number of series at our disposal was upwards of 250.

Tables.

Stage	Series	Length	Age	Blastomeres	Yolk
1	1		1 day 4 hrs.	2	
2	4		1 day 8 hrs.	6	
3	6		1 day 12 hrs.	12	Lower portion six surface grooves.
4	10		1 day 16 hrs.	20–24	Lower portion nine surface grooves; some reach center of egg.
5	12		2 days 2 hrs.	80–100	Lower portion twelve surface grooves; few reach center of egg.
6	18		2 days 7 hrs.	200–250	Lower surface as above excepting several grooves reach center of egg.
7	20		2 days 12 hrs.	300–600	Lower portion forty surface grooves; yolk much segmented.
8	25		6 days 16 hrs.		
9	28		10 days 16 hrs.		
10	33		13 days 3 hrs.		
11	35		14 days 4 hrs.		
12	40		14 days 19 hrs.		
13	48		15 days 10 hrs.		
14	55		16 days		
15	45		16 days 10 hrs.		
16	60		17 days		
17	73		17 days 17 hrs.		
18	85		18 days 15 hrs.		

Segmentation cavity	Blastopore. Anus	Archenteron. Enteron. Mesoderm. Chorda	Neurenteric canal	Nervous system. Optic vesicles. Somites	Stage
					1
					2
Present, large; oval single layer thick.					3
Large, roof in part two layers thick.					4
Segmentation cavity larger, roof in part two layers thick.					5
Segmentation cavity large, roof in part three layers thick.					6
Roof thinner, four layers around periphery, single layer in center.					7
Well formed.	Crescentic fissure on ventro-lateral portion of egg.	Archenteron just beginning.			8
Smaller.	Horse-shoe-shaped.	Archenteron extends over about 20°. Peristomal mesoblast present at dorsal lip of blastopore.			9
Smaller.	Circular.	Archenteron extends over about 45°. Dorsal wall posteriorly two layers. Peristomal mesoblast well defined in ventral lip of blastopore.			10
Quite small.	Small, circular, yolk plug externally visible.	Archenteron nearly complete; dorsal wall two layers anteriorly. Mesoblast around entire blastopore. Anlage of chorda.	Present.	Broad neural plate of thickened ectoblast; shallow neural groove.	11
Very small.	Very small, circular, yolk plug externally visible.	Archenteron very nearly complete; single layer in mid-longitudinal axis, this layer is beginning of chorda. Paraxial mesoblast co-extensive with chorda.	Present.	Low neural ridges present, deep neural groove.	12
Very small.	Elongated dorso-ventrally; yolk plug not externally visible.	Lateral walls of enteron three layers. Somatic and splanchnic layers well defined. Chorda well defined in head region.	Present, small.	Higher neural ridges, deep neural groove.	13
Very small.	Yolk plug not present.	Enteron single-layered dorsal wall; mesoblast extends ventrally over one-half of egg. Chorda narrower, thicker.	Tubular.	Prominent neural folds approaching each other, deep neural groove.	14
Not present (?).		Walls of enteron as above. Mesoblast extends to ventral portion of egg; well defined coelom. Chorda anteriorly a rod of cells.	Small canal.	High neural ridges approximating, but nowhere in contact.	15
Small, well defined.		Chorda oval or round in cross section throughout head and trunk regions, undifferentiated in tail.	Widely open.	Neural ridges meet in anterior head region. Beginning of optic vesicles.	16
(?)	Anus perforate.	Enteron as above. Mesoblast over ventral surface of egg. Chorda well defined in head and trunk regions.	Closed (?).	Neural ridges not closed in head region but coalesced in posterior trunk region and in tail. Optic vesicles better defined.	17
Present as large spaces among yolk cells.				Neural folds closed throughout entire length of embryo. Optic vesicles prominent, outer wall single layer of cells. Three or four somites.	18

Stage	Series	Length	Age	Body Form	Somites	Notochord	Nervous System	Eye	Ear
29	100 Trans. 101 Trans.	6 mm	20 days 10 hrs.						
29a		6½ mm							
30	102 Trans. 103 Sag.	7 mm	21 days 2 hrs.						
31	105 Trans.	8 mm	22 days 18 hrs.						
32	106 Trans. 107 Sag. 110 Front.	9 mm	23 days 10 hrs.						

							Single layer of long flat cells containing coarse yolk granules.	19
Anlage of nasal organ as thickening of internal layer of ectoderm.	Anlage of hypophysis as a long wedge-shaped mass of cells the inner end of which is 2 or 3 layers thick, outer end a single layer continuous with deeper layer of ectoderm. No capsule. Lies some distance from anterior end of notochord. Wall closely applied to single-layered wall of infundibulum. Yolk granules less numerous than in underlying entoderm.	Position of mouth indicated by slight thickening of ectoderm.	Gut extends from level of posterior portion of eyes to a point slightly beyond posterior end of embryo. Distended anteriorly to form branchial chamber; short protanal gut.	Anlage of thyroid as median outgrowth in anterior portion of floor of branchial cavity, in close relation to hyomandibular arch. No external indications of gill arches.	Prosencephalic ducts present as two short cells lying below straight the pharynx tubes.	Anlage of heart is form of a rod of and body and below between the veri-cells intereper-tral borders of the and with many pericardial cavities.	Thickened greatly over head. Over the long flat cells large oval cells.	20
Thickening of internal layer of ectoderm more pronounced and more definitely circumscribed.			Anlage of liver.	Thyroid evagination deeper. Outgrowth longer. Mandibular hyoid and common anlage of third, fourth and fifth arches visible externally.				21
Slight concavity in surface of ectoderm. Thick nasal organ equal to that of wall of forebrain. Narrow layer of mesenchyme between nasal organ and forebrain. External layer of ectoderm can be traced only to margin of nasal organ.	Mass of cells close under infundibulum. 3–5 cell layers thick. Slight indications of cavity. Has lost connection with ectoderm.	Surface invagination broad transverse groove.	Gut extends from anlage of mouth to junction of tail end of embryo with yolk; here it runs ventrad and extends somewhat further over the yolk than in the preceding stage. Extends laterally far beyond bounds of embryo at level of anterior portion of yolk. Diverticulum of gut just behind liver extends ventrally half way to ventral surface. Pharyngeal chamber large. Midgut narrow dorso-ventrally, wide laterally. No distinction between midgut and hindgut. Wall of pharynx and dorsal wall of midgut consist of single layer of columnar cells, heavily laden with yolk granules. Ventral wall of midgut irregular yolk cells. Liver pear-shaped evaginations.	Mandibular, hyoid, first branchial and fourth and fifth arches visible externally.	Prosencephalic ducts coiled, common anlage of open anteri-posteriorly. End about 10° of angular and End in mass inclose prom-inity to entoderm.	Pericardial cavity large. Heart has become a tube. Fusion-belong-tract. Division into auricular portion re-segment cognizable. Lateral blood vessels appearing, also branchial arches and sinus venosus.	Thickened over fore and mid-brain.	22

Stage	Series	Length	Age	Body-form	Somites	Notochord	Nervous System	Eye	Ear

Nose	Hypophysis	Mouth		Gills, Thyroid, Thymus, Trachea, Lung	Urinogenital System	Heart and Blood Vessels	Skin	Skin	Limbs	Stage
									Anlage of anterior limb visible.	23
										24
Nasal pit formed; organ oval in outline.									Anterior limb more prominent. Posterior limb buds not yet discernible.	25
										26

Stage	Series	Length	Age	Body Form	Somites	Notochord	Nervous System	Eye	Ear
27	127 Trans. 128 Sag. 130 Front.	14 mm	12 days 10 hrs						
28	131 Trans. 133 Sag. 134 Front.	15 mm	13½ days 13 hrs	Shows 32–34 myotomes.					
29	135 Trans. 138 Sag. 139 Front.	16 mm	15 days 16 hrs		Horizontal section shows 36–38 myotomes. In anterior trunk region much plate material is contralateral to level of prootic cleft. Cutis plate obscure.				

Nose	Hypophysis	Mouth	Digestive System, Liver, Pancreas, Spleen	Gills, Thyroid, Thymus, Trachea, Lung	Urinogenital System	Heart and Blood Vessels	Skin	Skeleton	Limbs	Stage

Stage	Series	Length	Age	Body Form	Somites	Notochord	Nervous System	Eye	Ear
30	130 Trans. 134 Sag. 135 Front.	17 mm	35 days		Horizontal section shows 40–42 myotomes.	Yolk not absorbed in anterior end. Notochord enlarged through trunk formed by between anterior end and hypophysis. Posteriorly against neural tube. Not segmented. Beginnings of neural arches in cartilage.	Pineum much less marked. Anterior end laterally, brain formed by developed; plexus mesenchyma. tends into lateral ventricles and diencephalon layers; contains some thick pigment. Antage of iris. Anterior commissure well defined, the commissure smaller, habenularis. Lamina cornu bellaris recognisable. Ganglion of trigeminal divided into two parts, now cleft. Spinoso-bulbar canal still presents minute ventrally, cells. ganglia very large; nerve roots well developed; nerve fibers appearing in ventral nerve roots. Antage of pia mater.	Optic cup flattened divided into two. Retina partition contains some spherical. inner wall line; dilated at this cavity except laterally a single layer of laminar dorso-ventrally, cells. Lens fibers forming.	Ear more extensive, membranous, surrounded by condensed cubo-saccular sertenly. Ductus endolymphaticus extends over lateral median cavity very of medulla. Lens larger over lateral distal end to. Anterior semicircular canal forming cartilage of laginarius and semicircular canals. Periotic capsule forming in condensed mesenchyma.
31	137 Trans. 139 Sag. 153 Front.	18 mm	40 days 20 hrs.	Trunk and tail slightly concave in profile. Cervical flexure still present. Cephalic flexure less marked. Tail very wide dorso-ventrally. Yolk more elongated, convex on ventral surface, slightly concave on dorsal margin. Pigmentation more pronounced. Mandibular arches coalesced. Boundaries of other arches no longer visible.	Horizontal section shows 44–46 myotomes.	Dorso-ventrally compressed. Yolk granules still present in peripheral portion. Neural arches better developed anteriorly.	Spinal cord shows well defined layer of fibers entirely around it. Layer thicker in ventral half. Neural portion. Transverse diameter of cord greater than dorso-ing. Around central walls. Lens sphere central a layer of cells radiation from canal. Nerve fibers numerous in ventral roots of spinal nerves.	Pigment increased in posterior layer of optic cup. Retina in three Optic stalks presents minute fibers beginning to be formed in posterior with fibers at posterior wall fills cavity; epithelium cubo-idal cells.	Periotic capsule beginning to appear on exterior side of ear.
32	159 Trans. 160 Sag. 162 Front.	19 mm	43 days		Horizontal section shows 48–50 myotomes.	Anterior end close to hypophysis; slightly bended. Portio anterior to heart much smaller than remaining position; greatly enlarged in level of anterior margin of yolk. Segmentation well marked in portion excepting meso-portion. Neural arches in anterior trunk region extend dorsally to level of middle of spinal cord.	Three layers well defined in walls of fore brain. Epiphysis a large oval better defined. well communication into body. third ventricle nearly extent. Walls of paraphysis irregular. Choroid plexus extending into ventricles. Lateral lobes of infundibulum extended to level of mid brain. Cerebral peduncles formed in ventro-lateral walls of mesencephalon.	Optic cup flattened laterally. Orbit better defined. Begins fibers in optic nerve better developed. Lens cells at nerve fibers numerous; not ragged concentrically in central portion.	Periotic capsule further down-medially.
33	161 Trans. 175 Sag. 171 Front.	20 mm	46 days 1 hrs.	Cephalic flexure reduced. Axis of tail coincides exactly with axis of trunk. Yolk reduced dorso-ventrally, ventrally, more pointed at anterior end. Dorsal and ventral surfaces convex. Pigment better developed.	Horizontal section shows some 50–55 myotomes. anterior trunk region muscle plate wider, extending centrally over margin of yolk. Septa thicker. Two thirds of trunk region indicated.	At level of anterior limb notochord much larger than in spinal cord, at level of posterior limb much smaller. Yolk not quite absorbed. Neural arches in anterior trunk extend dorsally to top of spinal cord, not united, beginning in posterior trunk region. In anterior caudal region indicated in procartilage.	Olfactory nerve connect with fibers from hemispheres.	Layers of retina well defined; pigment more dense. Lumen of optic stalk obliterated near brain wall; exceeds single minute details; obliterated near in center of lens socket. have nearly disappeared. Epithelium of lens a layer of cubo-idal cells. Fibers well defined, more numerous, concentrically arranged. Eye muscles better developed.	Utriculo-accessory in partition condense. tends well into vesicle. Anterior semicircular canal better defined. enclosed in other. Macula sacculus situated well defined. Periotic capsule surrounds ear except mesial surface.

Nose	Hypophysis	Mouth	Digestive System, Liver, Pancreas, Spleen	Gill, Thyroid, Thymus, Trachea, Lung	Urinogenital System	Heart and Blood Vessels	Skin	Skeleton	Limbs	Stage
										30
										31
										32
										33

Stage	Series	Length	Age	Body Form	Senses	Notochord	Nervous System	Eye	Ear
34	175 Trans. 176 Sag. 181 Front.	21 mm	39 days						
35	182 Trans. 183 Sag. 184 Front.	22 mm	52 days						
36	186 Trans. 188 Sag. 191 Front.	23 mm	55 days						

Nose	Hypophysis	Mouth	Digestive System, Liver, Pancreas, Spleen	Gills, Thyroid, Thymus, Trachea, Lung	Urino-genital System	Heart and Blood Vessels	Skin	Skeleton	Limbs	Stage
										31
										85
										9*

Stage	Series	Length	Age	Body Form	Somites	Notochord	Nervous System	Eye	Ear
37	192 Trans. 193 Sag. 197 Front.	24 mm	98 days			Anterior end in contact with parachordae; would but does not reach level of hypophysis. Constrictions well defined, also corpora; opposite bases of striatum (?). Pallium and central arches more subpellium differentiated marked anteriorly. Ossification well marked second growth. Haemal arches approximating ventrally in tail. Anlage of lateral basal process in cartilage in anterior portion of body.	Paraplysis with mass lateral diverticula; still opens into third ventricle. Eminentia Pallii medialis. ... Septum ependymale present. Beginning of taenia fornicis.		
38	198 Trans. 199 Sag. 200 Front.	25 mm	10 days	Flexures as above, yolk elongated, diameter reduced wider dorso-ventrally pying four and laterally fifths of Tail broader, space between pointed. Surface more chord and deeply pigmented; beads more marked dorsal ventrally to anterior and lower posterior limbs of yolk. possess Myoblasts digits. Pigment upper por beads more tion of layer sharply defined.	In anterior trunk region myotomes wider compressed in processes.	In anterior portion much compressed laterally, more compressed at bases of neural processes. At level of anterior limbs smaller than spinal cord; in trunk region and at level of posterior limbs large as cord; in tail about the diameter of cord. Yolk all absorbed. Haemal arches well developed, free coalesced. Ossification very considerable in neural arches and notochordal sheath. Lateral basal processes better developed throughout trunk region.			Ductus endolymphaticus extends over dorso-lateral walls of medulla; stalk very narrow but still widely open into saccule. Semicircular canals better developed; all partly enclosed in cartilage. Lagena & more extended evagination. Anlagen of pars neglecta and macula acustica neglecta. Macula acustica sacculi two layers of cells thick.
39	201 Trans. 204 Sag. 205 Front.	20 mm	114 days 4 hrs.	Compressed dorsoventrally in anterior part of body, laterally in posterior part. Anterior end surrounded by bone.	All divisions of brain clearly defined. Flexures laterally. extending from here spheres to medulla. Optic nerves very small, hollow near brain. Mid brain short, slightly arched. Ganglia habenulae large, symmetrical. Velum transversum contains numerous large blood vessels. Infundibulum wider than mid brain. Cerebellum extremely small. Spinal cord oval in section. White matter cells on outside thickens to lateral portions. Dorsal columns large, broader than anteriorly. Central canal very small nearly circular in section. Some yolk granules still present in cells of cord. Pia mater well developed.	Eye not so much flattened. Pictures laterally. Retina shows layer of rods and cones, outer nuclear layer, outer reticular layer, inner reticular layer, ganglionic layer. External limiting membrane well defined. Choroid better differentiated, pigment layer closely applied to pigmented layer of retina. Less spherical, surrounded by a capsule of a single layer of flattened cells. Fibres well developed. No nuclei in central epithelium in two layers. Eye muscles well developed.	

Nose	Hypophysis	Mouth	Digestive System, Liver, Pancreas, Spleen	Gills, Thyroid, Thymus, Trachea, Lung	Urino-genital System	Heart and Blood Vessels	Skin	Skeleton	Limbs	Stage
Wall of nasal duct consists of two layers of cuboidal cells.	Lies against floor of skull surrounded by a connective tissue capsule. More lobulated. Blood vessels growing into it.	Entire mouth open. Oesophagus open. Teeth projecting slightly into mouth cavity, ossified. Tongue better developed.	Glands in stomach well formed. In anterior half numerous, flask-shaped; in posterior half tubular. Both kinds unbranched. Yolk absorbed in epithelial cells of stomach also in hindgut; midgut cells still heavily laden. Oesophagus open. Liver extends farther caudally and dorsally. Hepatic cords more closely approximated. Yolk partially absorbed. Both dorsal and ventral pancreases larger, showing greater number of tubules. Spleen lies pressed against dorsal wall of body cavity; surrounded by thin layer of connective tissue; shows numerous dividing cells, much connective tissue.	Trachea longer reaching to caudal region of heart before dividing. Lungs longer, walls better defined, still very thick. Caudal end detached from intestinal wall except for short thin mesentery. Epithelium becoming flattened. Mesenchyma less compact. Lungs extend over anterior two thirds of liver laterally to dorso-lateral margins of liver. Posteriorly attached by band of mesenchyma to lateral margins of wall of stomach.	Pronephros reduced in size. Tubules cut off from coelom. Duct (meton.) opens into cloaca. Mesonephric tubules greatly convoluted; extends from 9" to 17" somite. Bladder comes off from cloaca near point where ureters enter. Sexual cells, a single row of cells containing large yolk granules. Opposite 15" segment.		Glands appearing in ventral region, anteriorly and posteriorly; numerous in lateral portion. Sense organs more numerous.	Supraoccipitals well formed in cartilage extending dorsally to level of spinal cord. Tibia and fibula chondrified; no tarsals or metatarsals. Ilium and pubis larger. Mesenchyme around nasal pits. Ossification beginning around margin of trabeculae. Parasphenoids forming as long plates on lower margin of basal plates extending forward to region of eye. Vomer ossified. Frontal ossified at sides and posterior end. Neural spines beginning to appear anteriorly. Indications of ossification in femur. Ossification beginning in neural processes and in haemal arches.		37
			Yolk nearly absorbed in liver.	Gills more prominent; filaments much branched. All clefts perforate excepting fifth. Lateral portion of thyroid contains small irregular vessels. Pleura proximally layer of flattened cells. Some mesenchyma between pleura and epithelium of lungs.			Oval cells and sense organs more numerous. Sensory hairs evident. Stratum corneum very flat cells. Stratum germinativum two layers.	Basilar plate continuous with capsular floor and trabeculae. Jugular foramen formed, elongated. Otic process of quadrate united with otic capsule. Antorbital process farther developed. Haemapophyses well developed; not united ventrally. Ossification in phalanges of anterior limb, also in iliac process.	Limbs better defined; anterior reach ventrally to border of yolk. Digits better defined in both.	38
Antero-lateral wall very thin. Lobulations prominent. Epithelium of dorsal wall ciliated. Olfactory nerve clearly defined.	Enlarged transversely. Divided into two portions composed of cells of different staining capacity, anterior part more deeply stained. Posterior part more lobulated and more vascular.	Unicellular glands in epithelium of mouth and tongue. Teeth extending into mouth cavity. Tongue well defined.	Unicellular glands present in mouth, pharynx, oesophagus and hindgut. Extreme anterior portion of stomach contains no glands. Pyloric glands much branched. Intestine lies in 4 to 5 transverse coils. Cavity of small intestine widely open. Muscular walls better developed. Epithelium consists of single layer of long columnar yolk-laden cells. Hindgut larger; yolk absorbed; anterior part shows evaginations (glands of Lieberkühn?). Liver lies closely in contact with dorsal wall of coelomic cavity. Yolk all absorbed. Gall bladder very large, surrounded by a layer of connective tissue, epithelium flat. Dorsal pancreas has grown caudally and contains 30 or more tubules; ventral contains about 12 tubules, is about ⅓ size of dorsal. Ducts of both lined with cubical epithelium.	Thymus consists of three spherical epithelial patches on each side. Two lie close together opposite posterior portion of ear. Anterior largest; posterior much the smallest. Trachea much larger, walls thinner, epithelium flatter. Lungs much larger, tapering caudally to end at level of posterior limbs; are attached to intestinal wall and to liver by short mesentery. Left lung longer than right and extends farther caudad.	Pronephros degenerating. Extends from middle of 5" segment to middle of 7", hanging down into body cavity. Duct very small. No openings into body cavity discernible. Mesonephros extends from 15" to 30" segments. Ducts open into cloaca. Bladder oval, thick-walled, much folded, opening into cloaca by short urethra, surrounded by connective tissue. Sexual glands two rods of large oval cells extending from 15" to 18" segment, just ventrad of mesonephros.	Auricles nearly same size in transverse section. Sinus venosus very large. Truncus arteriosus divided longitudinally. Trabeculae appearing in walls of auricles. Marked constriction between sinus and auricles.	Skin glands present in abdominal region. Stratum corneum a layer of squamous cells. Basement membrane dense.	Internasal cartilage connects trabeculae anteriorly. Pubic bones united in median line. Tarsals and metatarsals chondrified. Palatopterygoid and sphenoid ossified in posterior portion.		39

Stage	Series	Length	Age	Body Form	Somites	Notochord	Nervous System	Eye	Ear
40	206 Trans. 207 Sag. 208 Front.	17 mm	67 days			Chorda more compressed laterally at points where neural arches come in contact. In trunk region and in region of posterior limb transverse diameter equals that of spinal cord; in tail region diameter twice as great as that of cord. Beginning ossification in lateral arches.			Ear much extended antero-posteriorly, compressed dorso-ventrally. Ossification beginning in external portion of otic capsule.
41	210 Trans. 211 Sag. 212 Front.	28 mm	70 days 4 hrs.	General contour of body much changed, owing to rapid absorption of yolk. Tail broader dorso-ventrally. Constricted at level of posterior limbs. Yolk very long oval, flatter on ventral surface. Pigmentations extending down over one half of lateral surface of yolk.					
42	213 Trans. 214 Sag. 216 Front.	29 mm	74 days 12 hrs.						
43	217 Trans. 218 Sag. 219 Front.	30 mm	78 days	In anterior trunk region myotomes wider occupying more space between notochord and skin, extend ventrally to median line where they coalesce.	Much constricted in region where cartilages are in contact with body of vertebrae. Thickenings of sheath in long intervertebral disks better marked. Lie between layer of tissue between bone and sheath. Ossification beginning in lateral basal processes. Cartilaginous ribs present.	In brain and cord nearly in straight line. Groove prominent between diencephalon and mesencephalon. Deep groove in front of cerebellum. Pleuron large filling greater portion of 3rd ventricle. Cartilage extending into lateral ventricle and into 4th ventricle. Distal end of parapysis lies on level with epiphysis. All commissures better defined. Brain fills brain case at anterior end, not posteriorly. Spinal cord does not nearly fill canal. Dorsal columns of cord well defined; gray commissure present between them and gray matter of cord. Central canal very small, circular in section.		Endolymphatic sac wider, duct narrower. Cristae acusticae of canals formed. Macula acustica sacculi a thickened area of epithelium, three rows deep. Semicircular canals enclosed in cartilage. Ossification more extended on lateral surface of perotic capsule.

Nose	Hypophysis	Mouth	Digestive System, Liver, Pancreas, Spleen	Gills, Thyroid, Thymus, Trachea, Lung	Urinogenital System	Heart and Blood Vessels	Skin	Skeleton	Limbs	Stage
			Intestine crosses abdominal cavity 6—7 times. Dorsal portion of liver connected by narrow band with ventral portion. Hepatic cords closer together. Few hepatic ducts. Dorsal and ventral pancreases not in contact yet. Tubules more numerous in dorsal	Lungs reach to level of caudal end of liver.			Unicellular gland ginning in and sensory organs very numerous. Sensory hairs projecting beyond surface.	Ossification beginning in neural arches.		40
				Walls of trachea and lungs very thin; surrounded by mass of mesenchyma. Lungs flattened against stomach. Epithelium thinner; pleural covering very thin.				Ossification well defined in neural arches.		41
			Caudal portion of liver occupies over one half of coelomic cavity. Numerous hepatic ducts. Dorsal and ventral pancreases in contact. Dorsal much the larger. Spleen longer; compressed dorso-ventrally between intestine and dorsal body wall. Connective tissue more abundant.	Gills better developed, pigmented. Filaments greatly increased in number. Second, third and fourth gill clefts widely open to exterior; first and fifth imperforate.					Limbs longer, pigmented, both pairs functional. Four digits well defined on each.	42
Under lip marked by a very deep transverse groove. No shaped glands are larger. Tongue being cut off on ventral surface by deep lateral constriction.	Extreme anterior end of stomach contains no glands except unicellular. In anterior portion flask- and mucus cutaneous. Middle and posterior portion contain branched tubular glands. Intestinal epithelium better developed; still contains yolk granules. Histogical contains branched glands. Long process of liver extending posterior to gall bladder; another in median line extending over dorsal wall of gut.	Right lung lies against margin of liver; caudal end in contact with spleen.	Pronephric ducts small but still open. Mesonephros begin 3 or 4 segments behind pronephros and extends over about 15 segments. Bladder epithelium short. Mesonephric cells, or mucus cells in connection with excretive tissue. Walls much folded. Sexual glands extend from 13 to 20 segments. Yolk granules present in posterior portion.	Pericardium well defined, very thin, detached from muscular wall. Auricles widely in communication with each other. Blood passes into ventricle by single opening on left side. Valves present in epithelium short this opening. Semilunar valves between ventricle and conus appear as extensive proliferations of wall. Valves also present between conus and truncus.	Mucous glands larger, none so abundant. Unicellular more numerous.	All the cartilages of carpus, carpus and metatarsus and phalanges formed. Meckel's cartilage surrounded by bone. Premaxillae and frontals well ossified. Cartilage appearing in center of body of vertebrae.		43

Stage	Series	Length	Age	Body Form	Somites	Notochord	Nervous System	Eye	Ear
44	220 Trans. 221 Sag. 223 Front.	31 mm	82 days 12 hrs.						
45	224 Trans. 225 Sag. 226 Front.	32 mm	87 days			In caudal region larger than spinal cord. More constricted in centers of verte- brae. Intervertebral cartilages more pro- minent. Layer of bone thicker.			
46	227 Trans. 229 Sag. 230 Front.	33 mm	92 days						
47	231 Trans. 232 Sag.	34 mm	97 days	General form of body resembles adult. Ventral surface of head and trunk nearly in straight line. Anterior and posterior limbs project far below ventral surface of body. Pigmentation bands more precisely marked. Dorsal median dark band, below this a light band, then a broad dark band.		Bone surrounding anterior end thicker. Ribs better devel- oped; extend one half of distance to lateral surface of body. Neural arches at base more completely ossified.			

Nose	Hypophysis	Month	Digestive System, Liver, Pancreas, Spleen	Gills, Thyroid, Thymus, Trachea, Lung	Urino-genital System	Heart and Blood Vessels	Skin	Skeleton	Limbs	Stage
			Liver still larger in caudal portion occupying end of liver. Dorsal walls of body cavity. Tips of lungs in region subserial hepatic ducts rise to liver very much entering cystic duct. Gall bladder thin, ventral wall. Bladder removed some distance from ventral pancreas. Dorsal and ventral pancreases have united into a continuous mass.	Tip of lung at caudal end.						41
Compressed anteriorly, between infundibulum and floor of skull; broader posteriorly. Much lobulated.			Lumen of intestine larger; epithelium a single layer of poorly defined cells. Muscular wall very thin, a single layer of cells. Few hepatic ducts seen in body of liver; a few enter the cystic duct. Dorsal pancreas lobules compactly arranged.	Lungs very much larger, owing to thinness of walls and dorso-ventrally very much collapsed. Tips of lungs still show ament. Cartilages formed in walls of trachea. Epithelium of lateral walls of trachea much thicker than dorsal and ventral walls. Trachea in anterior portion shows columnar cells; ciliated? At level of heart epithelium flat.	Prosephros comparatively. extends over one segment; extends over about 10 segments.		Unicellular glands extend much more.			45
			Posterior portion of liver divided by transverse suture into smaller dorsal and larger ventral portion. Gall bladder very large; extends far caudal of liver; lined with flattened epithelium. Cystic duct larger and longer. Pancreas large extending around dorsal and lateral portion of intestine. Ducts open into cystic duct; show numerous branches with tubules emptying into them. Spleen larger, more vascular.		Second gonads extend far forward; begin to project down into body cavity; more distinctly separated from mesonephros.					46
			Intestine forms 6 or 7 transverse folds. Intestinal cavity very large. Muscular wall of midgut consists of two or more layers of cells. In hind gut unicellular glands very numerous. Dorsal and ventral pancreases more intimately united; dorsal larger and contains greater number of tubules. Spleen lies just to left of median line in concavity in dorsal wall of stomach.	Tracheal cartilages better developed at anterior end. Lungs or its obliterated; duct either side of stomach. Walls irregularly folded. Left lung still notably shorter than right and lying nearer mesial plane of body.	Prosephros smaller; beginning of many tubules degenerating. Mesonephros extends from 9° or 10° segment over about 43 segments. Müllerian duct not formed yet.			Scapula flattened laterally, broad antero-posteriorly. Coracoids meet in mid ventral line. Ossification beginning in pectoral and pelvic girdles, also in humerus and femur. Sphenoid ossified. Parietals ossified in median portion. Opisthoticum ossifying, also occipital arch.	Limbs much longer. Digits longer.	47

Stage	Series	Length	Age	Body Form	Somites	Notochord	Nervous System	Eye	Ear
36	211 Sag.	30 mm	110 days			Anterior end at level of hypophysis, sometimes beyond it. Intervertebral cartilages thicker. Bodies of vertebrae more constricted. Layer of bone around cord thicker.			
36a		35 mm							
39	234 Trans. 236 Sag. 237 Front.	39 mm	126 days	Has practically reached adult condition except in coloration. Some bands prominent as in preceding stage. Lateral band mottled.		Ossification very complete in neural spines and haemal processes in ribs, and in haemal arches; beginning in haemal spines.		Eye nearly spherical, slightly flattened laterally. Retina as in 26 mm excepting ganglionic layer one cell thick. Other layers together with external and internal limiting membrane sharply defined. Iris, choroid, cornea and sclerotic as in 26 mm. Lens capsule cells much more elongated horizontally.	

Nose	Hypophysis	Mouth	Digestive System, Liver, Pancreas, Spleen	Gills, Thyroid, Thymus, Trachea, Lung	Urinogenital System	Heart and Blood Vessels	Skin	Skeleton	Limbs/Stage

Some Variations in External Structures.

Before the individual variations are considered in detail it should be stated that a given nest contains but few eggs that are in precisely the same stage of development. The differences are most obvious in the early stages up to the closure of the neural fold. From the closure of the neural folds to the 30 mm larva the variations are not so pronounced, yet there are innumerable minor variations.

In the following descriptions the principal variations observed in the external features are first recorded, then those observed in the sections.

Cleavage.

In the cleavage stages, from the second on to late cleavage, there is so much variation in the position, extent and rate of progress of the various grooves that it is impossible to record them. Some of the variations have been described elsewhere (1904 b) by the senior author.

Gastrulation.

Some variation is found in the position of the first line of invagination which forms the dorsal lip of the blastopore. Its first appearance may be along a line equidistant from the equator and the vegetative pole or it may form nearer the equator and again sometimes nearer the vegetative pole. The first line of invagination may be nearly straight and again it may be decidedly crescentic. The maximal diameter of the yolk plug may equal one half the diameter of the egg. The closure of the blastopore usually occurs in about six days but it may close in five days.

Closure of neural folds.

Variations in the closure of the neural fold are frequently observed. The coalescence of the lateral portions of the fold usually begins at the posterior end. Sometimes they first coalesce along the middle portion of the embryo. Usually the transverse portion of the fold is continuous; at other times it shows a deep transverse groove which separates it into right and left halves. Cf. Figs. 18, 19, 20, 21. Wide variations exist in depth, width and extent of the neural groove.

Appearance of optic vesicles.

The optic vesicles are usually present before the complete closure of the neural fold. Sometimes they are present as disc-like depressions in the neural plate before the lateral portions of the fold have begun to coalesce; again, but rarely, they are not visible until the folds have closed.

Neuromeres (?).

There are frequently observed in the cephalic region well marked serial elevations and depressions along the inner margins of the lateral folds and across the neural plate. In other embryos there is not the slightest trace of either.

Somites.

In some embryos three myotomes are differentiated before the neural folds are closed, while in others there are no traces of myotomes until the folds are closed. The number of myotomes early becomes exceedingly variable in the tail. It is here impossible to count them accurately either in surface views or in sections. In the trunk i. e. between the limbs from the 15—16 mm larva up to the 39 mm they seem to be fairly constant numbering 18—20. In the tail however they are variable, so that in larvae of identical lengths there may be a variation of 1—5 myotomes. It should be emphasized that the number in the tail is determined with great difficulty since the most caudal are but slight thickenings in the mesoderm.

Lateral curvature of body.

In the embryos of 8—15 mm there is much variation in the lateral curvature of the body. In some nests as high as 80% of the embryos have the head and tail curved to the right. Other nests show a like percentage in which the head and tail are curved to the left. Out of 328 eggs, taken from five nests, 174 had the concave side on the right and 154 on the left.

External gills.

In most embryos of 9 mm (Fig. 25) the fourth and fifth arches are a common mass with no indications of the line of division; in some the line of invagination is distinct. In the later stages (e. g. 25 mm) there are usually five or six filaments on the middle gill bar, in others there are eight or nine. This variation is even more pronounced in the 26—39 mm larvae.

Limbs.

Some variations have been noted in the time of appearance of both the anterior and the posterior limb buds. The anlage of the anterior limb is usually discernible in the 11 mm stage (Fig. 27), but sometimes not until the larva measures 12 mm. The posterior limb buds are usually beginning in the 12 mm stage, sometimes are not present until the larva measures 13 mm. In the formation of the digits variations are found. Three are usually present on the anterior limbs in the 20—21 mm larva. In some three are not present until the larva is 23 mm long. The same variation is observed in the time of formation of the fourth digit on the anterior limb. Similar variations are found in the time of formation of the posterior digits.

Pigmentation.

Although little variation is observed in the position of the bands there is much variation in the degree of pigmentation. These differences are most pronounced in the larvae between 19—25 mm. In some the chromatophores are densely aggregated while in others they are sparsely scattered. In some (25 mm) they have extended over the dorsal portion of the yolk only, while in others they have extended over one half of the lateral surface of the yolk.

Variations in Internal Structures.

In the study of the variations of internal structures only those which are most obvious have been recorded.

Notochord.

There is considerable variation in the anterior extent of the notochord in nearly all stages. This variation is more obvious in the earlier than in the later stages. Usually a considerable tract of mesenchyma lies between the anterior end of the chorda and the hypophysis. Frequently this tract is short and sometimes only a narrow band.

Eye.

In sections the optic vesicles show the same variations which have been recorded under variations in external features. The appearance of pigment in the retina and iris shows considerable variation. In the 17 mm embryo series 140, 141, 145 show considerable pigment in the retina and its first appearance in the iris, while 140 shows the beginning of pigment in the retina and none in the iris. The time of first appearance of pigment in the choroid is likewise variable. In the 18 mm embryo series 152, 155 show considerable pigment, series 146, 151 show but little pigment, while in series 147, 149, 150 there is no pigment in the choroid.

Ear.

In the 6 mm embryo, series 100, there is a cup-shaped invagination of the inner layer of the ectoderm while in series 101 there is a vesicle. In the 7 mm embryo, series 102, shows the otic vesicle just closed, while in series 103 it is not quite closed. In the 18 mm larva, series 146 and 147, the endolymphatic duct extends over the medulla to about one half the distance to median line; in series 151 it extends over the lateral margin of the medulla; in series 149 it does not extend over the lateral margin of the medulla.

Nose.

In the 6 mm stage series 106 shows the anlage of the nose as a shallow invagination of the ectoderm. In series 107 there is no invagination. In 108 it is slightly cup-shaped. In series 109 thickened ectoderm, in series 110 thickened ectoderm.

Epiphysis.

In the 9 mm stage series 106, 104, 108 show the epiphysis to be a slight cup-shaped evagination in the dorsal wall of the fore-brain. In series 107, 110 the evagination is deeper and somewhat pear-shaped.

Paraphysis.

In the 11 mm embryos, series 116, the paraphysis is a very shallow evagination in the postero-dorsal wall of the telencephalon. In 117 the paraphysis is a deep pit-like evagination.

Hypophysis.

In the 9 mm embryo, series 106, 107, the anlage of the hypophysis is two layers of cells thick, in 108 it is but a single layer, while in 109 it is three or four layers thick.

Liver.

In the 9 mm embryo, series 106, 107, 108, 110 the anlage of the liver is shown as a wide evagination of the gut; while 109 shows two or three tubules.

Pancreas.

The dorsal pancreas seems to grow very slowly and to show much variation. Series 118 and 119 show the dorsal pancreas as a slight evagination of the dorsal wall of the gut. In series 120 it is in about the same stage. In series 125 the evagination is vesicular. In the 13 mm embryo series 121, there is a cup-shaped evagination. In 122, 124 the evagination is vesicular. In the 14 mm stage it varies from cup-shaped evagination in series 127, 130 to vesicular in 129.

Spleen.

In the 15 mm embryo, series 131, 134 show the anlage of the spleen as a small mass of mesenchymal cells in the dorsal mesentery of stomach. In series 132 there is no indication of spleen. In 133 there is a well defined mass of mesenchymal cells.

Thymus.

In the 16 mm embryo, series 137, 138, 139 the thymus begins as a small mass of cells in close proximity to the 3rd branchial cleft. In 136 the thymus anlage is not yet present.

Trachea—Lungs.

In the 10 mm embryo series 111, 112 show the beginning of the trachea as a short groove in the ventral wall of the pharynx. Series 113 shows a deeper vesicular evagination.

Bibliography.

The literature comprises the papers on those forms which belong to the Ichthyoidea (CLAUS) including the Perennibranchiata (*Proteus, Siren, Necturus*) and those of the Caducibranchiata belonging to the family Derotremata (*Amphiuma, Menopoma, Cryptobranchus*).

It includes all (?) the papers on the embryology, nearly all those on the anatomy and histology. Most of the systematic, zoological and paleontological papers are also given.

A. Alphabetical arrangement of titles according to authors.

1878 ABBOTT, CHARLES C., Our largest Salamander (*Menopoma alleghaniensis*). Science Gossip, 1878, p. 271—272.
1878 ALBRECHT, PAUL, Ueber einen Processus olentoideus des Atlas bei den urodelen Amphibien. Centralbl. f. d. med. Wissensch., 1878, p. 578—579. Nachtrag, p. 705—709.
1903 ALCOCK, E. P., On certain features of the cranial anatomy of *Bdellostoma dombeyi*. Anat. Anz., Bd. 23, 1903, p. 259—281, 321—327.
1904 ANGEL, P., Étude du développement des glandes de la peau des Batraciens et en particulier de la Salamandre terrestre. Arch. de Biol., T. 19, 1904, p. 257—289, 2 pl.
1892 ABERNCROS, O. A., Zur Kenntnis des sympathischen Nervensystems der urodelen Amphibien. Zool. Jahrb., Bd. 5, 1892, p. 184—210, 4 pl.
1896 ANDRES, ANGELO, La Salamandra gigantesca del Giappone (*Megalobatrachus maximus* Boie.). Atti. Soc. Ital. Sc. nat., Vol. 35, 1896, p. 203—218, 1 fig.
1898 ANDREWS, E. A., Filose activity in metazoan eggs. Zool. Bull. Boston, Vol. 2, 1898, p. 1—13, 5 fig.
1898a ANDREWS, E. A., Some activities of polar bodies. Ann. Mag. Nat. Hist., Vol. 1, 1898, p. 109—116, 5 fig.
1819 BAIRD, SPENCER F., Revision of the North American tailed batrachians with descriptions of new genera and species. Journ. Acad. Nat. Sci. Philad., Ser. 2, Vol. 1, 1819, p. 281—294.
1826 BARNES, DANIEL H., An arrangement of the genera of the Batrachian animals with a description of the more remarkable species including a monograph of the doubtful reptiles. Amer. Journ. Sci. and Arts, Ser. 1, Vol. 11, 1826, p. 268—297.
1827 BARNES, DANIEL H., Note on the doubtful reptiles. Amer. Journ. Sci. and Arts, Ser. 1, Vol. 13, 1827, p. 66—70.
1807 BARTON, BENJ. SMITH, (Scientific notes). Philad. Med. and Physical Journ., Suppl. 2, 1807.
1808 BARTON, BENJ. SMITH, Notices of Siren lacertina and of another species of the same genus. Philadelphia 1808. 8°.
1812 BARTON, BENJ. SMITH, Some account of the Siren lacertina and other species of the same genus of amphibious animals. Philadelphia 1812. 8°.
1812a BARTON, BENJ. SMITH, A memoir concerning an animal of the class of Reptilia or Amphibia, which is known in the United States by the names of Alligator and Hellbender. Philadelphia 1812. 8°. 1 pl.
1885 BAUR, G., Einige Bemerkungen über die Ossifikation der „langen" Knochen. Zool. Anz., Bd. 8, 1885, p. 580—581.
1888 BAUR, G., Beiträge zur Morphologie des Carpus und Tarsus der Vertebraten. Teil 1. Batrachia. Jena 1888.
1889 BAUR, G., On the morphology of the vertebrate skull. 1. The otic elements. Journ. Morphol., Vol. 3, 1889, p. 367—474.
1891 BAUR, G., The pelvis of the Testudinata, with notes on the evolution of the pelvis in general. Journ. Morphol., Vol. 4, 1891, p. 345—359.
1894 BAWDEN, H. H., The nose and Jacobson's organ with especial reference to the Amphibia. Journ. Comp. Neurol. and Psychol., Vol. 4, 1894, p. 117—152, 8 pl.
1878 BEALE, LIONEL S., The microscope in medicine, 4. ed. London 1878.
1799 BEAUVOIS, Memoir on a new species of Siren (*operculata*). Trans. Amer. Phil. Soc., Vol. 4, 1799, p. 277—281.
1902 BECKER, VICTOR, Untersuchungen an der Mundschleimhaut von *Cryptobranchus japonicus*. Diss. Phil., Berlin 1902, p. 1xt.
1901 BERNARD, F. K., Normally inequal growth as a possible cause of death. Nature, Vol. 68, 1903, p. 407.
1901 BERNARD, F. K., On some points in the anatomy chiefly of the heart and vascystem of the Japanese Salamander, *Megalobatrachus japonicus*. Proc. Zool. Soc. London, Vol. 2, 1904, p. 298—315, 7 fig.
1899 BIRR, TH., Die Accommodation des Auges bei den Amphibien. Arch. Ges. Phys. Pflüger, Bd. 73, 1899, p. 481—554, 14 Fig.

1835 Bell, Thomas, Amphibia. In Todd's Cyclopaedia of anatomy and physiology, Vol. 1, London 1835/1836, p. 90—107.
1907 Bender, Otto, Die Histologie des Spritzloches der Selachier und der Paukenhöhle der Amphibien, Sauropsiden und Säugetiere und Grund ihrer Innervation. Anat. Anz., Ergänzungsheft zu Bd. 30, 1907, p. 38—44.
1907a Bender, Otto, Die Schleimhautnerven des Facialis, Glossopharyngeus und Vagus. Studien zur Morphologie des Mittelohres und der benachbarten Kopfregion der Wirbeltiere. Denksche. Med.-naturw. Ges. Jena, Bd. 7, 1907, p. 541—651.
1880 Bensley, R. R., The oesophageal glands of Urodela. Biol. Bull., Vol. 2, 1880, p. 87—101, 8 Fig.
1898 Drew, John M., A comparison of the phagocytic action of leucocytes in Amphibia and Mammalia. Trans. Amer. Micr. Soc., Vol. 19, 1898, p. 105—116, 5 pl.
1860 Bertholdi, Beta H., Der Riesensalamander von Japan (Megalobatrachus maximus). Natur, Bd. 9, 1860, p. 211—212.
1846 Dugès, F. H., Vergleichend-anatomische und histologische Untersuchungen über die männlichen Geschlechts- und Harnwerkzeuge der nackten Amphibien. Dorpat 1846, p. 1—74.
1889 Doliewski, N. F., Physiologische Bemerkung über den Riesensalamander (Russian). Arb. d. Naturf. Ges. Charkow, Bd. 15, 1889, p. 173—206.
1871 Blanchard, Émile, On a new gigantic Salamander (Sieboldia davidiana Blanch.) from Western China. Ann. Mag. Nat. Hist., Ser. 4, Vol. 8, 1871, p. 232—244.
1871a Blanchard, Émile, Note sur une nouvelle Salamandre gigantesque (Sieboldia davidiana Blanch.) de la Chine occidentale. Compt. Rend. Acad. Sc. Paris, 1871, p. 79—80.
1891 Blanchard, Rapi, Évacuation de noyaux cellulaires simulant une helminthiase et une coccidiose (Proteus anguineus). Bull. Soc. Zool. France, T. 16, 1891, p. 22—28.
1874 Doyère, E., Magenepithel und Magendrüsen der Batrachier. Inaug.-Diss. Königsberg, 1874.
1881 Ihlis, J. E. V., Ueber den cotus arteriosus und die Arterienbogen der Amphibien. Morphol. Jahrb., Bd. 7, 1881, p. 488—572, 3 Taf.
1826 Boie, H., Merkmale einiger japanischer Lurche. Isis, 1826, p. 203—216.
1899 Bolau, Henk, Glandula thyreoidea und Glandula Thymus der Amphibien. Zool. Jahrb., Abt. f. Anat., Bd. 12, 1899, p. 657—710, 11 Fig.
1832 Bonaparte, Carlo Luciano, Fauna Italica. Roma 1832—1841, T. 2, Tab. 131.
1838 Bonaparte, Carlo Luciano, Iconografia della Fauna Italica. Bibl. Ital., 93, 1838.
1839 Bonaparte, Carlo Luciano, Amphibia europaea ad systema nostrum vertebratorum ordinata. Torino 1839, 4°, p. 72.
1876 Boettger, Oskar, Ueber die inneren Kiemenöffnungen bei jungen Exemplaren des japanischen Riesensalamhs. Zool. Garten, Bd. 17, 1876, p. 432—435.
1882 Boulenger, G. A., Catalogue of the Batrachia gradientia s. caudata and Batrachia apoda in the Collection of the British Museum. 2. ed. London 1882, 8°, p. 8—127, 9 pl.
1907 Braucht, A., Recherches sur l'anagenese de la tête chez les Amphibiens. Arch. de Biol., T. 23, 1907, p. 165—257.
1907 Braus, H., Ueber Frühanlagen der Nebelarmmuskeln bei Amphibien und ihre allgemeinere Bedeutung. Anat. Anz., Erg.-Heft, Bd. 30, 1907, p. 192—219.
1900 Bridge, T. W., The air-bladder and its relationship with the auditory organ in Nocturus bornensis. Journ. Linn. Soc. Zool., Vol. 27, 1900, p. 503—540, 2 pl.
1908 Broili, J. F., Zur Fortpflanzung des japanischen Riesensalamanders (Megalobatrachus maximus). Natur und Haus, Bd. 12, 1908.
1901 Broman, J., Notiz über das "Habstück" der Spermien von Pelobates fuscus nebst kritischen Bemerkungen über die Nomenklatur der Spermienschwanzfäden. Anat. Anz., Bd. 20, 1901, p. 347—351, 3 Fig.
1904 Broman, J., Die Entwicklungsgeschichte der Bursa omentalis und ähnliche Recessbildungen bei den Wirbeltieren. Wiesbaden 1904, p. 614, 650, Fig. 20 Pl.
1902 Brunn, O., Ueber den Ursprung des Kopfskelettes bei Necturus. Leipzig 1902, 8°. Also Morph. Jahrb., Bd. 29, 1901, p. 582—613.
1873 Bugnion, E., Recherches sur les organes sensitifs qui se trouvent dans l'épiderme du Proteus et de l'Axolotl. Bull. Soc. Vaudoise de Sci. nat. Lausanne, T. 12, 1873.
1874 Bugnion, E., Sur Proteus anguineus. Bull. Soc. Vaudoise Sci. nat. Lausanne, T. 12, 1874, p. 193—194.
1874a Bugnion, E., Sur le système nerveux du Proteus anguineus. Bull. Soc. Vaudoise Sci. nat. Lausanne, T. 13, 1874, p. 441—448.
1896 Buretz, H. C., A contribution to the study of variation. Skeletal variation of Necturus maculatus Raf. Journ. Morphol., Vol. 12, 1896, p. 455—464, 2 pl.
1904 Bussy, L. P. de, Eerste ontwikkelingsstadien van Megalobatrachus maximus Schlegel. Amsterdam 1904.
1905 Bussy, L. P. de, Die ersten Entwicklungsstadien des Megalobatrachus maximus Schlegel. Zool. Anz., Bd. 38, 1905.
1883 Camerano, Lorenzo, Intorno alla neotenia ed allo sviluppo degli Anfibi. Atti. Acc. Tor., T. 19, 1883, p. 84—93.
1883a Camerano, Lorenzo, Ricerche intorno alla vita branchiale degli Anfibi. Zool. Anz., Bd. 6, 1883, p. 685—687.

1808 CUVIER, PIERS., De Dugon du comte de Buffon et de la Sirena lacertina de chevalier LINNÆUS. Oeuvres, T. 3, Paris 1808.

1886 CALMELS, A. J., Untersuchungen über Gliedmassenreste bei Schlangen. Bihang till K. Svenska Vet. Akad. Handlingar, Bd. 11, 1886.

1886 CARNOY, A. J., Die Ganglienzellen des Bulbus anteriores und der Kammerspitze beim Salamander (Necturus maculatus). Arch. f. d. ges. Physiol., Bd. 102, 1906, p. 51—60, 3 Fig.

1878 CHAPMAN, HENRY C., Notes on the Amphiuma. Proc. Acad. Nat. Sci. Philadelphia 1879, p. 144—145.

1880 CHAPMAN, HENRY C., Observations on the Japanese Salamander Cryptobranchus maximus (SCHLEGEL). Proc. Acad. Nat. Sci. Philad., 1882, p. 227—243, 2 pl.

1888 CHAUVIN, MARIE VON, Vorläufige Mitteilung über die Fortpflanzung des Proteus anguineus. Zool. Anz., Bd. 6, 1883, p. 330—332.

1883a CHAUVIN, MARIE VON, Die Art der Fortpflanzung des Proteus anguineus. Zeitschr. wiss. Zool., Bd. 38, 1883, p. 671—685, 1 Taf.

1884 CHAUVIN, MARIE VON, Ueber die Färbung des Männchens von Proteus anguineus. Naturforscher, Bd. 16, 1884, p. 480.

1840 CHIAJE, STEFANO DELLE, Ricerche anatomico-biologiche sul Proteo serpentino. Napoli 1840, 21 p. 1°, 5 pl.

1897 CHRONCHAMPSKY, Boris, Entstehung der Milz und des dorsalen Pankreas beim Necturus. Compt. Rend. XII. Congrès intern. zool. Moscou, Vol. 2, 1897, p. 115—130.

1900 CHRONCHAMPSKY, Boris, Die Entstehung der Milz, Leber, Gallenblase, Bauchspeicheldrüse und des Pfortadersystems bei den verschiedenen Abteilungen der Wirbeltiere. Anat. Hefte, Bd. 13, 1900, p. 363—423, 85 Fig.

1882 CLAUS, C., Lehrbuch der Zoologie. 2 Bde. 4. ed. Leipzig 1882.

1889 CLAYPOLE, Edith J., An investigation of the blood of Necturus and Cryptobranchus. Proc. Amer. Micr. Soc., Vol. 15, 1893, p. 39—76, 6 pl.

1896 CLAYPOLE, Edith J., Notes on the comparative histology of blood and muscle. Trans. Amer. Micr. Soc., Vol. 18, 1896, p. 49—70, 5 pl.

1895 CREMENS, P., Die äusseren Kiemen der Wirbeltiere. Anat. Hefte, Bd. 5, 1895, p. 51—156, 4 Taf., 5 Fig.

1821 CONFIGLIACHI, P., und RUSCONI, M., Observations on the natural history and structure of the Proteus anguinus. Edinb. Phil. Journ., Vol. 4, 1821, p. 388—406, and Vol. 5, p. 84—112.

1885 COPE, E. D., Amphibia. In Standard Natural History, Vol. 3, Boston 1885, p. 303—347.

1881 COPE, E. D., The retrograde metamorphosis of Siren. Amer. Nat., Vol. 19, 1886, p. 1226—1227.

1889a COPE, E. D., On the structure and affinities of the Amphiumidae. Amer. Philos. Soc., 1886.

1889 COPE, E. D., The Batrachia of North America. Bull. 34 U. S. Nat. Mus., Washington 1889.

1889a COPE, E. D., On the relations of the hyoid and otic elements of the skeleton in the Batrachia. Journ. Morphol., Vol. 2, 1889, p. 297—310, 3 pl.

1895 COPE, E. D., A careless writer on Amphiuma. Amer. Nat., Vol. 29, 1895, p. 1103—1110.

1860 CULP, Edward, Note on the blood-corpuscles of the Japanese gigantic Salamander (Sieboldia maxima). Proc. Zool. Soc. London, 1860, Part 28, p. 203—205.

1800 CUVIER, G., Sur le Siren lacertina. Bull. des Sci. Soc. Philom., An. 8, T. 2, p. 106—107.

1827 CUVIER, G., Sur le genre de Reptiles batraciens nommé Amphiuma et sur une nouvelle espèce de cet genre (A. tridactylum). Mem. du Muséum, T. 14, 1827, p. 1—14, 2 pl.

1831 CUVIER, G., Règne animale. Translated by EDWARD GRIFFITH and others. London 1831, Vol. 9, p. 412.

1853 BARTON, Ino. C., Some account of the Proteus anguinus. Amer. Journ. Sci., Ser. 2, Vol. 15, 1853, p. 378—388.

1894 DAVISON, ALVIN, Amphiuma tridactyla. Princeton Coll. Bull., 1894.

1894a DAVISON, ALVIN, The arrangement of muscular fibres in Amphiuma tridactyla. Anat. Anz., Bd. 9, 1894, p. 332—336.

1896 DAVISON, ALVIN, A contribution to the anatomy and phylogeny of Amphiuma means (GARDEN). Journ. Morphol., Vol. 11, 1895, p. 37—414, 2 pl.

1896 DAVISON, ALVIN, The testacular apparatus of Amphiuma. Amer. Nat., Vol. 30, 1896, p. 681—682, 4 fig.

1897 DAVISON, ALVIN, A preliminary contribution to the development of the vertebral column and its appendages. Anat. Anz., Bd. 14, 1897, p. 6—12, 7 Fig.

1834 DIJEN, J. van, Over de rijbelingsche takken van de zwervende zenuw (Nervus vagus), van den Proteus anguineus. Tijdschr. voor natuurl. Geschied., Bd. 1, 1834, p. 112—129.

1842 DE KAY, James E., Zoology of New York. Part 3. Reptiles and Amphibia. In Natural History of New York. Albany 1842. 4°.

1850 DUMERIL, B., Natural history of the Amphiumidae with remarks on Crocodilian hibernation and instinct. New Orleans Med. and Surg. Journ., Vol. 16, 1850, p. 14—37.

1894 DRÜNER, L., Studien zur Anatomie der Zungenbein-, Kiemenbogen- und Kehlkopfmuskeln der Urodelen. 1. Thl. Zool. Jahrb., Bd. 15, 1901, p. 435—622, 7 Taf.

1903 DRÜNER, L., Ueber die Muskulatur des Visceralskelets der Urodelen. Anat. Anz., Bd. 23, 1903, p. 545—571, 16 Fig.

1903 DUTTENIES, ANNA, Sur le tissu lymphoïde du rein du Proteus anguineus LAUR. Note préliminaire. Compt. Rend. Soc. Biol. Paris, T. 55, 1903, p. 1091—1092.

1889 DUBOIS, RAPH., Sur la perception des radiations lumineuses par la peau chez les Protées aveugles des grottes de la Carniole. Compt. Rend. Ac. Sc. Paris, T. 110, 1890, p. 358—361.

1834 DUGÈS, A., Recherches sur l'ostéologie et la myologie des Batraciens. Paris 1834.

1834 DUMÉRIL, AND. MAR. CONST., et BIBRON, G., Erpétologie générale, ou histoire naturelle complète des Reptiles. 8 Vol., Paris 1834—1854. 8°. Avec pl.

1847 DUVERNOY, G. L., Ueber die Geschlechts- und die Harnorgane der daumendgeschwänzten Batrachier. Froriep Not. 1847, No. 203.

1859 ECKER, A., Notiz über den Olm Proteus anguineus LAURENTI. Correspond.-Bl. d. Naturf. Ver. Riga, Bd. 11, 1859, p. 123—124.

1882 ECKER, E. C., and PUTNAM, F. C., Catalogue of the Reptiles and Batrachians New York State. N. Y. State Museum Bull., Vol. 10, 1882, No. 51, p. 353—358, 410—414, 1 pl.

1864 EHRMANN, LUDWIG, Vorlesungen über den Bau der nervösen Zentralorgane des Menschen und der Tiere. 2 Bde. 7. ed. Leipzig 1904, 1908.

1862 EHRENBERG, CHR., Ueber die Nahrung des Proteus anguineus. Sitzungsber. d. Ges. naturf. Freunde Berlin, 1862.

1867 EHRENBERG, CHR., Ueber den seit 7 Jahren lebend gehaltenen Proteus anguineus. Sitzungsber. d. Ges. naturf. Freunde Berlin, 1867, p. 1.

1868 EHRENBERG, CHR., Ueber den seit 9 Jahre in der Gefangenschaft gehaltenen Proteus anguineus. Sitzungsber. d. Ges. naturf. Freunde Berlin, 1868, p. 14—15.

1870 EHRENBERG, CHR., Ueber den in Gefangenschaft gehaltenen Proteus anguineus. Sitzungsber. d. Ges. naturf. Freunde Berlin, 1870, p. 2.

1872 EHRENBERG, CHR., Ueber den seit 1859 lebend gehaltenen Proteus anguineus, und über den seit 1859 beobachteten Triton lacustris. Sitzungsber. d. Ges. naturf. Freunde Berlin, 1872, p. 17—18.

1874 EHRENBERG, CHR., Ueber den seit 1859 lebend erhaltenen Proteus anguineus. Sitzungsber. d. Ges. naturf. Freunde Berlin, 1874, p. 9—10.

1899 EIGENMANN, C. H., A case of convergence. Science, Vol. 9, 1899, p. 280—282, 3 fig.

1899 EIGENMANN, C. H., The eyes of the blind vertebrates of North America. 2. The eyes of Typhlomolge rathbuni STEJ. Trans. Amer. Micr. Soc., Vol. 21, 1900, p. 49—60.

1900a EIGENMANN, C. H., Degeneration in the eyes of cold-blooded vertebrates of the North American caves. Science, Vol. 11, 1900, p. 492—503, 14 fig.

1899 EIGENMANN, C. H., and DENNY, W. A., Eyes of blind vertebrates of North America. 3. The structure and outogenic degeneration of the eyes of the Missouri Cave Salamander. Biol. Bull., Vol. 2, 1900, p. 33—41, 9 fig.

1900b EIGENMANN, C. H., and DENNY, W. A., Eyes of Cave Salamander (Typhlotriton spelaeus). Proc. Indiana Acad. Sci., 1899, p. 252—255.

1899 EISEN, GUSTAV, Blood plates of the human blood with notes on the erythrocytes of Amphiuma and Necturus. Journ. Morphol., Vol. 15, 1899, p. 633.

1875 EISMANN, GUST., Der Adelsberger Olm Proteus anguineus in Gefangenschaft. Zool. Garten, Bd. 16, 1875, p. 394—396.

1766 ELLIS, JOHN, An account of an Amphibious Bipes (Siren lacertina). Phil. Trans. Roy. Soc. London, Vol. 56, 1766, p. 189—192, 1 pl.

1906 EMERSON, E. T., General anatomy of Typhlomolge rathbuni. Proc. Boston Soc. Nat. Hist., Vol. 32, 1906, p. 43—76, pl. 1.

1897 FRAAS, C., Die fossilen Reste von Archegosaurus und Eryops und ihre Bedeutung für die Morphologie des Gliedmaßenskelets. Anat. Anz., Bd. 14, 1897, p. 201—208, 7 Fig.

1863 FRAAS, JOSEPH, Beobachtungen an Amphibien in der Gefangenschaft. Verh. d. Zool.-bot. Ges. Wien. Bd. 13, 1863, p. 129—132.

1864 FRAAS, JOSEPH, Die Amphibien der österr. Monarchie. Mit Anlehnung der Beobachtungen die an den in der Gefangenschaft gehaltenen Arten gemacht wurden. Verh. d. zool. bot. Ges. Wien, Bd. 14, 1864, p. 697—720.

1877 FRAAS, JOSEPH, Ueber die Lebensweise von Siren lacertina in der Gefangenschaft. Verh. d. Zool.-bot. Ges. Wien, Bd. 26, 1877, p. 114—116.

1893 EYCLESHYMER, ALBERT C., Development of the optic vesicles in Amphibia. Journ. Morphol., Vol. 8, 1893, p. 189—194.

1895 EYCLESHYMER, ALBERT C., Early development of Amblystoma. Journ. Morphol., Vol. 10, 1895, p. 343—400.

1902 EYCLESHYMER, ALBERT C., The formation of the embryo of Necturus, with remarks on the theory of concrescence. Anat. Anz., Bd. 21, 1902, p. 311—353.

1902a EYCLESHYMER, ALBERT C., Nuclear changes in the striated muscle cell of Necturus. Anat. Anz., Bd. 21, 1902, p. 379—385.

1904 EYCLESHYMER, ALBERT C., The cytoplasmic and nuclear changes in the striated muscle cell of Necturus. Amer. Journ. Anat., Vol. 3, 1904, p. 285—310, 4 pl.

56   Namensinden zur Entwicklungsgeschichte der Wirbeltiere.

1901a EYCLESHYMER, ALBERT C., Bilateral symmetry in the egg of Necturus. Anat. Anz., Bd. 25, 1904, p. 230—230, 47 Fig.

1906 EYCLESHYMER, ALBERT C., The habits of Necturus maculosus. Amer. Nat., Vol. 40, 1906, p. 123—136.

1906a EYCLESHYMER, ALBERT C., The growth and regeneration of the gills in the young Necturus. Biol. Bull., Vol. 10, 1906, p. 171—175, 1 fig.

1906b EYCLESHYMER, ALBERT C., The development of the chromatophores in Necturus. Amer. Journ. Anat., Vol. 5, 1906, p. 309—330, 7 fig.

1907 EYCLESHYMER, ALBERT, The Closing of Wounds in the Larval Necturus. Amer. Journ. of Anat., Vol. VII, p. 317—326.

1906 FAVARO, GIUS., Ricerche anatomo-embriologiche intorno alla circolazione caudale ed ai cuori linfatici posteriori degli Anuri, con particolare riguardo agli Urodeli. Atti Accad. Sc. Venet. Trent. Padova, Anno 8, 1906, p. 122—100, 20 fig.

1906 FELIX, W., and BÜHLER, A., Die Entwicklung der Harn- und Geschlechtsorgane. Handbuch der Entwicklungslehre der Wirbeltiere von OSKAR HERTWIG, Jena 1906, Bd. 3, p. 81.

1894 FIXEN, HERBERT B., Sur le développement des organes extérieurs chez l'Amphiuma. Compt. Rend. de l'Acad. Sci. Paris, T. 118, 1894, p. 1221—1224.

1896 FELIX, HERBERT B., Bemerkungen über die Entwicklung der Wirbelsäule bei den Amphibien; nebst Schilderung eines abnormen Wickelsegmentes. Morphol. Jahrb., Bd. 22, 1895, p. 340—356.

1843 FISCHER, J. G., Amphibiorum nudorum neurologia. Berolini 1843.

1864 FISCHER, J. G., Anatomische Abhandlungen über die Perennibranchiaten und Derotremen. Hamburg 1864. 4°. p. 170, 6 pl.

1900 FISCHER, SIGWART, Ueber Proteus anguineus L. Verh. Schweiz. Naturf. Ges., 1900, p. 76.

1826 FITZINGER, LEOPOLD J., Neue Klassifikation der Reptilien. Wien 1826. 4°.

1850 FITZINGER, LEOPOLD J., Ueber den Proteus anguineus. Wien. Sitzungsber., Math.-nat., Kl., Bd. 5, 1850, p. 291—303.

1902 FLEISSIG, JULIUS, Zur Anatomie der Nasenhöhle von Cryptobranchus japonicus. Anat. Anz., Bd. 25, 1902, p. 48—54, 5 Fig.

1861 FRANTZIUS, G. Ritter von, Notiz über den Olm. Verh. d. Zool.-bot. Ges. Wien, Bd. 12, 1861, p. 29.

1882 FREUD, WM., Vitality of the Mud Puppy (Menopoma). Amer. Nat., Vol. 16, 1882, p. 325—326.

1846 FREYER, HEINRICH, Ueber eine neue Art von Hypochthon (Proteus). Arch. f. Naturgesch., Bd. 12, 1846, p. 289—293.

1894 FÜRBRINGER, F., Bericht über eine zur Untersuchung der Entwicklung von Axolotl, Lepidosteus und Necturus unternommene Reise nach Nordamerika. Sitzungsber. d. Akad. d. Wiss. Berlin, Bd. 49, 1894, p. 1057—1070.

1877 FÜRBRINGER, M., Zur Entwicklung der Amphibienniere. Heidelberg 1877.

1906 FÜRBRINGER, M., Zur vergleichenden Anatomie des Brustschulterapparates und der Schultermuskeln. Jena. Zeitschr., Bd. 36, 1902, p. 289—736, 5 Taf., 23 Fig.

1895 GADOW, HANS, On the evolution of the vertebral column of Amphibia and Amniota. Proc. Roy. Soc. London, Vol. 56, 1895, p. 257—259.

1901 GADOW, HANS, Amphibia and Reptilia. Cambridge Natural History, Vol. 8, London 1901.

1882 GAGE, SIMON HENRY, Observations on the fat cells and connective tissue corpuscles of Necturus (Menobranchus). Proc. Amer. Soc. Micr. Buffalo, Vol. 5, 1882, p. 109—126, 1 pl.

1885 GAGE, SIMON HENRY, Notes on the epithelium lining the mouth of Necturus and Menopoma. Proc. Amer. Soc. Micr. Buffalo, Vol. 8, 1885, p. 126.

1885a GAGE, SIMON HENRY, Notes on the blood corpuscles of Necturus. Proc. Amer. Soc. Micr. Buffalo, Vol. 8, 1885, p. 126.

1890 GAGE, SUSANNA PHELPS, The intramuscular endings of fibres in the skeletal muscles of the domestic and laboratory animals. Proc. Amer. Soc. Micr., Vol. 13, 1890, p. 132—138, 4 pl.

1890 GAGE, SIMON HENRY, and GAGE, SUSANNA PHELPS, Changes in the ciliated areas of the alimentary canal of the amphibia and the relation to the mode of respiration. Proc. Amer. Assoc. Adv. Sci., Vol. 39, 1890, p. 337—338.

1897 GARBOTTI, GINO, Studio morfologico e citologico della volta del diencefalo in alcuni Vertebrati. Riv. Pat. nerv. e ment. Firenze, Vol. 2, 1897, p. 484—517, 20 fig.

1888 GÜNTHER, J. H., On a new species of Menobranchus (lateralis var. Latastei). Proc. Canad. Inst., Ser. 3, Vol. 5, 1888, p. 278—289.

1901 GAUPP, ERNST, Das Hypobranchialskelet der Wirbeltiere. Ergeb. Anat. u. Entwickl., Bd. 14, 1901, p. 808—1048, 46 Fig.

1906 GAUPP, ERNST, Die Entwicklung des Kopfskeletes. Handbuch der Entwicklungslehre der Wirbeltiere von OSKAR HERTWIG. Bd. 3, Jena 1906.

1881 GIARD, A. J. C., Notice sur la grande Salamandre du Japon. Cryptobranchus japonicus v. BOIVEN. Nouv. Arch. Mus. Paris, T. 2, 1881, p. 273—299, 1 pl.

1884 GIARD, A. J. C., Notice sur la grande Salamandre du Japon. Nouv. Arch. du Mus., T. 5, 1884, p. 273—290.

1902 GILGENREINER, CARL, Untersuchungen zur vergleichenden Anatomie der Wirbelsäule bei Amphibien und Reptilien. Leipzig 1902. 4°. p. 22.

1898 Gegenbaur, Carl, Vergleichende Anatomie der Wirbeltiere. 2 Bde. Leipzig 1898.

1905 Geyer, Hans, Der Schlammteufel (Cryptobranchus alleghaniensis). Natur u. Haus, Bd. 13, 1905, p. 353—356, 1 Fig.

1850 Gibbes, Lewis R., On a new species of Menobranchus from South Carolina (M. punctatus). Proc. Amer. Assoc. Adv. Sci., Vol. 3, 1850, p. 159.

1853 Gibbes, Lewis R., Description of Menobranchus punctatus. Boston Journ. Nat. Hist., Vol. 6, 1853, p. 369—573, Fig. 1.

1806 Goeze, Carl Chrn., Gemeinnützliche systematische Naturgeschichte für gebildete Leser nach dem Linnéschen Natursystem. (German, Latin and French text.) 4 Teile. Mannheim 1801—1829.

1891 Goeppert, E., Die Entwicklung und das spätere Verhalten des Pancreas der Amphibien. Morphol. Jahrb., Bd. 17, 1891, p. 100—122, 4 Taf.

1894 Goeppert, E., Die Kehlkopfmuskulatur der Amphibien. Morphol. Jahrb., Bd. 22, 1894, p. 1—78.

1895 Goeppert, E., Zur Kenntnis der Amphibienrippen. Morphol. Jahrb., Bd. 22, 1895, p. 411—448, 5 Fig.

1896 Goeppert, E., Die Morphologie der Amphibienrippen. Festschr. f. Gegenbaur, Bd. 1, 1896, p. 235—455, 2 Taf. 10 Fig.

1898 Goeppert, E., Der Kehlkopf der Amphibien und Reptilien. Morphol. Jahrb., Bd. 26, 1898, p. 282, 2 Taf., 5 Fig.

1896 Goeppert, E., Die Entwicklung des Mundes und der Mundhöhle mit Drüsen und Zunge; die Entwicklung der Schwimmblase, der Lunge und des Kehlkopfes der Wirbeltiere. Handbuch der Entwicklungslehre der Wirbeltiere von Oskar Hertwig, Jena 1906, Bd. 2, p. 1—108.

1825 Gray, John Edward, A synopsis of the genera of Reptiles and Amphibia, with a description of some new species. Thomson Ann. Philos., Vol. 10, 1825, p. 193—217.

1837 Gray, John Edward, On the genus Necturus or Menobranchus with an account of its skull and teeth. Proc. Zool. Soc. London, Vol. 25, 1837, p. 61—64.

1873 Gray, John Edward, On a Salamander (Sieboldia) from Shanghai. Ann. Mag. Nat. Hist., Ser. 4, Vol. 12, 1873, p. 138.

1896 Green, Isabella M., The peritoneal epithelium in Amphibia. Amer. Nat., Vol. 30, 1896, p. 911—915.

1897 Green, Isabella M., The peritoneal epithelium of some Ichsen Amphibia (Necturus, Amblystoma, Desmognathus and Diemyctylus). Trans. Amer. Micr. Soc., Vol. 18, 1897, p. 76—108, 5 pl.

1876 Grote, A. R., On casting the skin in Menopoma (Cryptobranchus) alleghaniensis. Amer. Journ. Sci., Ser. 3, Vol. 12, 1876, p. 472.

1876a Grote, A. R., Note on Menopoma alleghaniensis of Harlan. Amer. Journ. Sci., Vol. 12, 1876.

1896 Grube, E., Ueber die Nahrung des Proteus anguinus. Jahrb. d. Schles. Ges. f. vaterl. Kultur, 1896, p. 63—64.

1856 Gueerne, Jules de, Mort de la grande Salamandre du Japon (du Muséum d'Histoire naturelle de Paris). Etangs et Rivières, Vol. 10, 1856, p. 267.

1873 Gulliver, George, Measurements of the red blood corpuscles of Batrachians. Proc. Zool. Soc. London, 1873, p. 162—165.

1904 Hall, R. W., The development of the mesonephros and the Müllerian ducts in Amphibia. Bull. Mus. Comp. Zool. Harvard, Vol. 45, 1904, p. 31—125, 8 pl., 1 fig.

1856 Hallowell, Edw., Pseudotriton marginatus and flavissimus ns. spp. Urodel. from Georgia. Proc. Acad. Nat. Sci. Philad., Vol. 8, 1856, p. 130—131.

1858 Hallowell, Edw., On the caducibranchiate Urodele Batrachians. Journ. Acad. Nat. Sci. Philad., Ser. 2, Vol. 3, 1858, p. 337.

1892 Hargitt, Charles W., On some habits of Amphiuma means. Science, Vol. 20, 1892, p. 159.

1823 Harlan, Richard, Description of a Batrachian animal in a living state (Amphiuma means). Journ. Acad. Nat. Sci. Philad., Vol. 3, 1823, p. 54—59.

1824 Harlan, Richard, Observations on the genus Salamandra with the anatomy of the Salamandra gigantea (Barton) or S. alleghaniensis (Michaux) and two new genera proposed. Ann. Lyc. Nat. Hist., Vol. 1, 1824, Pt. 2, p. 222—234.

1824a Harlan, Richard, Dissection of a Batrachian animal in a living state (Amphiuma means). Philos. Mag., Vol. 63, 1824, p. 323—329, 1 fig.

1825 Harlan, Richard, Ueber Amphiuma means. Frorr. Not., Bd. 14, 1826, p. 8—10, Fig.

1826a Harlan, Richard, Further observations on the Amphiuma means. Ann. Lyc. Nat. Hist. New York, Vol. 1, 1826, p. 269.

1827 Harlan, Richard, American herpetology, or Genera of the North American Reptilia; with a synopsis of the species. Journ. Acad. Nat. Sci. Philad., Vol. 5, 1827, Pt. 2, p. 324.

1835 Harlan, Richard, Medical and physical researches; or Original memoirs. Philad. 1835, p. 164—176, fig.

1858 Hartog, Pieter, Note sur les corpuscules sanguins du Cryptobranchus japonicus. Verslagen en Mededeelingen der Koninklijke Akad. van Wetensch., Afdeeling Natuurkunde, Vol. 7, 1858, p. 368—372, 1 pl.

1871 Harting, Pieter, Nieuwe Reuzen-Salamander. Album der Natur, 1871.

1888 Hay, O. P., Observations on Amphiuma and its young. Amer. Nat., Vol. 22, 1888, p. 315—321.

1889 Hay, O. P., On the structure of the skull of the larva of Amphiuma. Proc. Amer. Assoc. Adv. Sci., 1889, p. 286.

1890 Hay, O. P., The skeletal anatomy of Amphiuma during its earlier stages. Journ. Morphol., Vol. 4, 1890, p. 11—34, 1 pl.

1892 Hay, O. P., The Batrachians and Reptiles of Indiana. Indianapolis 1892. 8°. p. 204, 3 pl.

1893 Heidenhain, M., Die Hautdrüsen der Amphibien. Sitz.-Ber. Physik.-med. Ges. Würzburg, 1893, No. 4, p. 52—64.

1887 Heidenhain, M., Ueber die Zentralkapseln und Pseudochromosomen in den Samenzellen von Proteus, sowie über ihr Verhältnis zu den Idiosomen, Chondromiten und Archoplasmaschleifen. Anat. Anz., Bd. 18, 1900, p. 513—598, 8 Fig.

1907 Heidenhain, M., Plasma und Zelle. Handbuch der Anatomie des Menschen von Karl von Bardeleben, Bd. 6, Abt. 2, Jena 1907.

1897 Reighard, Clara, A peculiar pelvic attachment in Necturus maculatus. Biol. Bull., Vol. 12, 1907, p. 375—377.

1893 Herrick, C. L., The callosum and hippocampal region in marsupial and lower brains. Journ. Comp. Neurol. and Psychol., Vol. 3, 1893, p. 176—182.

1894 Herrick, C. Judson, The cranial nerves of Amblystoma. Journ. Comp. Neurol. and Psychol., Vol. 4, 1894, p. 193—207.

1899 Herrick, C. Judson, The nervus terminalis (nerve of Pinkus) in the Frog. Journ. Comp. Neurol. and Psychol., Vol. 10, 1899, p. 203—209, 1 Fig.

1906 Hertwig, Oscar, Allgemeine Biologie. Jena 1906. 8°.

1889 Hess, C., Beschreibung des Auges von Talpa europaea und von Proteus anguineus. Arch. f. Ophthalmol., Bd. 35, 1889, p. 1—19, 1 Taf.

1900 Hilton-Taber, O., Ueber das Gehirn von Proteus anguineus. Arch. mikr. Anat. u. Entw., Bd. 72, 1900, p. 719—731.

1906 Hochstetter, F., Die Entwicklung des Blutgefäßsystems. Handbuch der Entwicklungslehre der Wirbeltiere von Oskar Hertwig, Jena 1906, Bd. 3, p. 21—166.

1838 Hoeven, J. van der, [Letter to Mr. Owen]. Proc. Zool. Soc. London, 1838, p. 23.

1838a Hoeven, J. van der, Jets over den grooten zoogenoemden Salamander van Japan. Met afbeeldingen van schedels en eene nieuwe aanteekening van de Menopoma van Harlan. Tijds-chr. Natuurl. Geschied. Physiol., Vol. 4, 1838, p. 375—386, 2 pl.

1838b Hoeven, J. van der, Sur une nouvelle espèce de Cryptobranchus du Japon. Bull. Sci. phys. natur. Néerlando, Leiden 1838, p. 90—91.

1839 Hoeven, J. van der, Note sur une nouvelle espèce de Cryptobranchus. Ann. Sci. nat., T. 11, 1839, p. 63—64.

1841 Hoeven, J. van der, Groote bloodschijfjes bij Cryptobranchus japonicus. Tijdschr. voor Natuurl. Geschied., Vol. 8, 1841, p. 270—272.

1846 Hoeven, J. van der, Note sur le carpe et le tarse du Cryptobranchus Japonicus. Arch. néerl. Sc. exact. et nat., T. 1, 1866, p. 321—327.

1866a Hoeven, J. van der, Considérations sur le genre Menobranchus et sur ses affinités naturelles. Arch. néerl. Sc. exact. et nat., T. 1, 1866, p. 1—16.

1866a Hoeven, J. van der, Notes on the genus Menobranchus and its natural affinities. Ann. Mag. Nat. Hist., Vol. 18, 1866, p. 363—375.

1867 Hoeven, J. van der, Ontleed- en Dierkundige Bijdragen tot de Kennis van Menobranchus. Arch. néerl. Sc. exact. et nat., T. 2, 1867, p. 289.

1867a Hoeven, J. van der, Les globules du sang du Menobranchus. Arch. néerl. Sc. exact. et nat., T. 2, 1867, p. 288.

1867b Hoeven, J. van der, Ontleed- en dierkundige bijdragen tot de Kennis der Menobranchus, den Proteus der meren van Noord-Amerika. Leiden 1867. 8 pl.

1867c Hoeven, J. van der, Over Menobranchus. Proc. verh. Vergad. Kon. Akad. Wetensch. Amsterdam, Afd. Natuurk., 1867, p. 3.

1877 Hoffmann, C. K., Amphibien. Bronn's Klassen und Ordnungen des Tierreiches, Leipzig 1877, Bd. 6, Abt. 2.

1902 Hoffmann, C. K., Zur Entwicklungsgeschichte des Sympathicus. Verh. Akad. Amsterdam, Vol. 8, 1902, No. 3, p. 161, 4 pl.

1875 Holbrook, John Edwards, North American Herpetology; or a Description of the Reptiles inhabiting the United States. Vol. 5, Ed. Pl. Philad., 1836—43, roy. 4°.

1903 Holmgren, E., Zur Kenntnis der cylindrischen Epithelzellen. Arch. mikr. Anat. u. Entw., Bd. 65, 1903, p. 280—297, 2 Taf., 3 Fig.

1903 Hornbeck, Henry S., Muscular and skeletal elements in Spelerpes longicaudus. Ohio Naturalist, Vol. 3, 1903, p. 379—393, fig. 1.

1908 Howard, Arthur Day, The visual cells in vertebrates, chiefly in Necturus. Journ. Morphol., Vol. 19, 1908, p. 561—629, 5 pl.

1888 Howes, G. B., and Davies, A. M., Observations upon the morphology and genesis of supernumerary phalanges with especial reference to those of Amphibia. Proc. Zool. Soc. London, 1888.

1893 Howes, G. B., Notes on variation and development of the vertebral and limb skeleton of the Amphibia. Proc. Zool. Soc. London, 1893, Pt. 2, p. 268—278, 15 fig.

1806 Humboldt, Alexander von, Recueil d'observations de zoologie et d'anatomie comparée, 2 T., Paris 1805—32, 55 pl.

1872 Humphry, G. M., The muscles and nerves of the Cryptobranchus japonicus. Journ. Anat. and Physiol., Vol. 6, 1872, p. 1—61, 4 pl.

1866 Hyrtl, John, Supplement to (John Ellis's) account etc. being the anatomic description of the said animal [Siren lacertina]. Phil. Trans. Roy. Soc. London, Vol. 56, 1766, p. 307—310.

1871 Huxley, Thomas Henry, On the structure of the skull and heart of Menobranchus. Proc. Zool. Soc. London, 1874, p. 186—204, 1 pl.

1875 Huxley, Thomas Henry, Amphibia. In: Encyclopaedia Brittanica. 9. ed., Vol. 1, 1875, p. 758—771.

1850 Hyrtl, J., Bemerkungen über den Proteus anguineus. Wien. Sitz.-Ber., math.-nat. Kl., Bd. 5, 1850, p. 300.

1865 Hyrtl, J., Cryptobranchus japonicus. Schediasma anatomicum. Vindobonae 1865.

1865a Hyrtl, J., Kurze Inhaltsanzeige einer im nächsten Jahr zu veröffentlichenden Abhandlung über die Anatomie des Riesensalamanders. Wien. Sitz.-Ber., math.-nat. Kl., Bd. 1, 1865, p. 48—49.

1890 Ishikawa, C., Zur Entwicklungsgeschichte von Cryptobranchus japonicus. Toga Gakuge-Zashi, 1890.

1892 Ishikawa, C., Ueber den Riesensalamander Japans. Mitt. Ges. f. Naturk. Ostasiens, Bd. 9, 1892, p. 79—91.

1893 Ishikawa, C., Distribution, oviposition, habits etc. of the Japanese giant-salamander (Japanese). Tokyo Teishitsu Hakubutsukwan, Vol. 2, 1893, p. 27.

1904 Ishikawa, C., Beiträge zur Kenntnis des Riesensalamanders (Megalobatrachus maximus Schl.). Proc. Depart. Nat. Hist. Tokyo, Imperial Museum, Vol. 1, 1904.

1905 Ishikawa, C., The gastrulation of the gigantic Salamander Megalobatrachus maximus. Zool. Magazine, Vol. 17, 1905.

1908 Ishikawa, C., Ueber den Riesensalamander Japans (Cryptobranchus japonicus). Entwicklung der äußeren Körperform. Mitt. Ges. f. Naturk. Ostasiens, 1908, 22 p., 3 Taf., 3 Fig.

1823 James, Edwin, Account of an expedition from Pittsburg to the Rocky Mountains 1819. From the notes of Major Long, Mr. T. Say and others. 2 Vols. Philadelphia 1823. 8°. With atlas.

1897 Julliette, Sarah Ely, Preliminary note upon the cytology of the brains of some Amphibians. I. Necturus. Journ. Comp. Neurol. and Psychol., Vol. 7, 1897, p. 146—151.

1876 Johnston, Christopher, Observations upon spermatozoa of Amphiuma tridactylum. Monthly Micr. Journ., Vol. 16, 1876, p. 61—63, 1 pl.

1903 Johnston, J. B., Movements of the cerebro-spinal fluid in Cryptobranchus. Science, N. S., Vol. 17, 1903, p. 530.

1906 Johnston, J. B., The nervous system of Vertebrates. Philadelphia 1906.

1905 Joseph, H., [Randreifen der roten Blutkörperchen der Proteus]. Verh. Anat. Ges. 19. Vers., 1905, p. 229—230.

1905a Joseph, H., Ueber die Zentralkörper der Nervenzelle. Verh. Anat. Ges. 19. Vers., 1905, p. 178—187, 16 Fig.

1905 Kammerer, P., Ueber die Abhängigkeit des Regenerationsvermögens der Amphibienlarven von Alter, Entwicklungsstadium und spezifischer Größe. Arch. Entw.-Mech., Bd. 19, 1905, p. 148—180.

1905a Kammerer, P., Die angebliche Ausnahme von der Regenerationsunfähigkeit bei den Amphibien. Centralbl. Physiol., Bd. 19, 1905, p. 684—687.

1907 Kammerer, P., Die Fortpflanzung des Grottenolmes (Proteus anguineus Laurenti). Verh. k. k. Zool.-bot. Ges. Wien, Bd. 57, 1907, p. 277—292.

1903 Kuipers, P. N. van, Eieren en spermién van Megalobatrachus maximus Schl. Tijdsch. Ned. Dierk. Ver., Vol. 8, 1903.

1901 Kopsch, F., Die Gastrulation und die Keimblätterbildung der Wirbeltiere. Ergeb. Anat. Entw., Bd. 10, 1901, p. 1002—1119, 43 Fig.

1906 Keibel, F., Die Entwicklung der äußeren Körperform der Wirbeltierembryonen, insbesondere der menschlichen Embryonen aus den ersten 2 Monaten. Handbuch der Entwicklungslehre der Wirbeltiere von Oskar Hertwig. Bd. 1, 1906, p. 174.

1903 Kerbert, C., Eieren van Megalobatrachus maximus Schl. Tijdschr. d. Ned. Dierk Ver., Vol. 8, 1903.

1904 Kerbert, C., Zur Fortpflanzung von Megalobatrachus maximus Schl. Zool. Anz., Bd. 27, 1904, p. 305—329, 1 Fig.

1905 Kerbert, C., Ueber die Eier und Larven von Megalobatrachus maximus Schl. Compt. Rend. 6 Cong. Internat. Zool. Bâle, 1905, p. 289—294.

1894 Kingsbury, B. F., The histological structure of the enteron of Necturus maculatus. Proc. Amer. Micr. Soc. Ithaca, Vol. 16, 1894, p. 19—64, 8 pl.

1895 Kingsbury, B. F., The spermatheca and methods of fertilization in some American Newts and Salamanders. Trans. Amer. Micr. Soc., Vol. 17, 1895, p. 261—304, 4 pl.

1895a Kingsbury, B. F., On the brain of Necturus maculatus. Journ. Comp. Neurol. and Psychol., Vol. 5, 1895, p. 139—205, 3 pl.

1895b Kingsbury, B. F., The lateral line system of sense organs in some American Amphibia and comparisons with Dipnoans. Proc. Amer. Micr. Soc., Vol. 17, 1895, p. 115—150, 5 pl.

1903 Kingsbury, B. F., Columella auris and Nervus facialis in the Urodela. Journ. Comp. Neurol. and Psychol., Vol. 3, 1903, p. 313—334.

1904 Kingsbury, B. F., The rank of Necturus among tailed Batrachia. Biol. Bull., Vol. 6, 1904, p. 67—74.

1908 Kingsbury, B. F. and Reed, H. D., The columella auris in Amphibia. Anat. Record, Vol. 2, 1908, p. 81—94.

1892 Kingslett, J. S., The head of an embryo Amphiuma. Amer. Nat., Vol. 26, 1892, p. 671—680.

1889 Kingslett, J. S., and Kurroer, W. H., The ossicula auditus and mammalian ancestry. Amer. Nat., Vol. 33, 1899, p. 219—290, 3 fig.

1900 Kingslett, J. S., The ossicula auditus. Tufts Coll. Stud., 1900, No. 6, p. 293—274, with pl.

1902 Kingslett, J. S., The systematic position of the Caecilians. Tufts Coll. Stud., 1902, No. 7, p. 293—321, 3 pl.

1883 Klaussner, F., Das Rückenmark des Proteus anguineus L. München 1883.

1886 Knaffl, E., Das Binnenrohr Organ. Morphol. Jahrb., Bd. 11, 1886, p. 489—552.

1856 Kneeland, Samuel, Letter relating to a supposed new species of Siredon from Lake Superior. Proc. Boston Soc. Nat. Hist., Vol. 6, 1856—59, p. 152—154.

1856a Kneeland, Samuel, On the breathing apparatus of the Menobranchus. Proc. Boston Soc. Nat. Hist., Vol. 6, 1856—59, p. 428—430.

1858 Kneeland, Samuel, [Notes on Menobranchus] Proc. Boston Soc. Nat. Hist., Vol. 6, 1858, p. 152—154, 218, 371—373.

1868 Kneeland, Samuel, The largest blood discs known. Singular capture of a Canadian Reptile. Menobranchus lateralis. Intellectual Observer, Vol. 12, 1868, p. 194—200.

1895 Knoll, Ph., Ueber die Blutkörperchen bei wechselwarmen Wirbeltieren. Sitz.-Ber. Kais. Akad. Wiss. Wien, Bd. 105, 1896, p. 35—66, 3 Taf., 4 Fig.

1889 Kölliker, Albrecht von, Handbuch der Gewebelehre des Menschen. Fortgesetzt durch V. von Ebner. 3 Bde. Leipzig 1889.

1903 Kohnstein, Hans, Zur Morphologie und Physiologie des Gefäßsystems am Respirationstrakt. Anat. Hefte, Bd. 22, 1903, p. 307—375, 4 Taf., 2 Fig.

1903a Kohnstein, Hans, Die Funktion der Muskulatur der Amphibienlunge. 1. Anatomischer Teil. Arch. ges. Physiol., Bd. 95, 1903, p. 616—624.

1889 Kohl, C., Einige Notizen über das Auge von Talpa europaea und Proteus anguineus. Zool. Anz., Bd. 12, 1889, p. 383—386, 405—408.

1891 Kohl, C., Vorläufige Mitteilung über das Auge von Proteus anguineus. Zool. Anz., Bd. 14, 1891, p. 93—96.

1851 Kornhuber, G. A., Bemerkung über das Vorkommen des Olm (Proteus anguineus Laur.). Verh. d. Ver. f. Naturk. Preßburg, Bd. 8, 1851, p. 55—57, 1 Fig.

1861 Kraus, L. M., Weitere Mitteilung über das angebliche Lebendiggebären des Proteus. Verh. d. Zool.-bot. Ges. Wien, Bd. 12, 1861, p. 37.

1879 Krug, Ueber das häutige Labyrinth der Amphibien. Arch. mikr. Anat. u. Entw., Bd. 17, 1879, p. 479—550.

1886 Kraus, W., Neue Untersuchungen über die motorische Nervenendigung. Zeitschr. f. Biol., Bd. 23, 1886, p. 148, 16 Taf.

1903 Kunstorn, Ueber Ringgefäße in der Epidermis von Cryptobranchus japonicus. Mitt. d. Med. Ges. Tokio, Bd. 17, 1903.

1906 Keibel, K. von, Die Morphogenie des Zentralnervensystems. Handbuch der Entwicklungslehre der Wirbeltiere von Oskar Hertwig, Bd. 2, Jena 1906, p. 1—304.

1807 Lacépède, M. de, Sur une espèce de quadrupède ovipare non encore décrite, Proteus tetradactylus [Menobranchus lateralis] Ann. Mus. Hist. nat. Paris, T. 10, 1807, p. 230—235, 1 pl.

1807a Lacépède, M. de, Sur une espèce de Protée ou Salamandre à quatre doigts à toutes les pattes. Nouv. Bull. Sci. Soc. Philom., T. 1, 1807, p. 64.

1906 Lange, Dan de, De Kieuwblaadvorming van Megalobatrachus maximus Schl. Amsterdam 1906.

1907 Lange, Dan de, Die Keimblätterbildung des Megalobatrachus maximus (Schl.). Anat. Hefte, Bd. 32, 1907, p. 311—477, 4 Taf.

1873 Langerhans, Paul, Notiz zur Anatomie des Amphiüenhornes. Zeitschr. wiss. Zool., Bd. 23, 1873, p. 475—480, 1 Fig.

1906 Lankes, K., Necturus maculatus Raf. in bayerischen Gewässern. Bl. Aquarien-Kunde, Magdeburg, Bd. 17, 1906, p. 160—161.

1902 Lauber, H., Anatomische Untersuchung des Auges von Cryptobranchus japonicus. Anat. Hefte, Bd. 20, 1902, p. 4—17.

1826 Le Conte, John L., Description of a new species of Siren [striata] with some observations on animals of a similar nature. Ann. Lyc. Nat. Hist. New York, Vol. 1, 1826, p. 52—58.

1828 Le Conte, John L., Description of the Siren lacertina n. sp. Ann. Lyc. Nat. Hist. New York, Vol. 2, 1828, p. 133—134.

1849 Le Conte, John L., On the habits of Amphibia in a state of captivity. Proc. Amer. Assoc. Adv. Sci., 1849, p. 194—195.

1821 Leuckart, F. S., Einiges über die fischartigen Amphibien. Isis, Litter. Anz., 1821, p. 260, 1 Taf.

1840 Leuckart, F. S., Ueber den Genus Cryptobranchus. Frorieps Notizen, Bd. 13, 1840, p. 19—20.

1897 Levi, Giu., Ricerche citologiche comparate sulla cellula nervosa dei Vertebrati. Riv. Pat. nerv. ment. Firenze, Vol. 2, 1897, p. 193—225, 244—255, 1 tab., 5 fig.

1904 Lewis, Fred T., The question of sinusoids. Anat. Anz., Bd. 25, 1904, p. 261—279, 10 Fig.

1853 Leydig, F., Anatomisch-histologische Untersuchungen über Fische und Reptilien. Berlin 1853. 4°. p. 6 u. 120. 4 Taf.

1868 Leydig, F., Ueber Organe eines sechsten Sinnes. Zugleich ein Beitrag zur Kenntnis des feineren Baues der Haut bei Amphibien und Fischen. Nova Acta Acad. Caes. Leopold. Dresden 1868.

1873 Leydig, F., Ueber die allgemeinen Bedeckungen der Amphibien. Arch. mikr. Anat. u. Entw., Bd. 9, 1873, p. 753—794, 4 Taf.

1876 Leydig, F., Ueber die allgemeinen Bedeckungen der Amphibien. Arch. mikr. Anat. u. Entw., Bd. 12, 1876, p. 119—242.

1876a Leydig, F., Ueber die Schwanzblase, Taschörperchen und Endorgane des Nerven bei Batrachiern. Arch. mikr. Anat. u. Entw., Bd. 12, 1876, p. 523.

1876b Leydig, F., Die Hautdecke und Hautsinnesorgane der Urodelen. Morphol. Jahrb., Bd. 2, 1876, p. 287—318, 4 Taf.

1898 Leydig, F., Vaskularisiertes Epithel. Arch. mikr. Anat. u. Entw., Bd. 52, 1898, p. 152—455.

1766 Linnaeus, C., Siren lacertina. Upsala 1766. 4°. p. 15, 1 pl.

1896 Ijima, W. Akimoto, The gigantic Japanese Salamander (Sieboldia maxima). Science Gossip, 1896, p. 130—131.

1906 Lord, Clarence, Some cellular changes in the primary optic vesicles of Necturus. Journ. Comp. Neurol. and Psychol., Vol. 15, 1905, p. 452—464, 1 pl.

1890 Lucas, F. A., The sacrum of Menopoma. Amer. Nat., Vol. 20, 1886, p. 564—565.

1887 Macallum, A. B., The termination of nerves in the liver. Quart. Journ. Micr. Sci., Vol. 27, 1887, p. 439—460.

1887a Macallum, A. B., On the nuclei of the striated muscle fibre in Necturus (Menobranchus lateralis). Quart. Journ. Micr. Sci., N. S., Vol. 26, 1887, p. 439—460.

1892 Macallum, A. B., Studies on the blood of Amphibia. Trans. Canad. Inst., Vol. 2, 1892, Part 2, p. 221—260, 4 pl.

1908 Mc Gill, Caroline, The fibroglia fibres of Necturus. Anat. Record, Vol. 2, 1908, p. 146.

1908a Mc Gill, Caroline, Fibroglia fibres in the intestinal wall of Necturus and their relation to myofibrils. Internat. Monatsschr. f. Anat. u. Physiol., Bd. 25, 1908, p. 90—98, 1 pl.

1896 Mc Gregor, J. Howard, An embryo of Cryptobranchus. Proc. N. Y. Acad. Sci., 1896.

1896a Mc Gregor, J. Howard, Preliminary notes on the cranial nerves of Cryptobranchus alleghaniensis. Journ. Comp. Neurol. and Psychol., Vol. 6, 1896, p. 45—53.

1899 Mc Gregor, J. Howard, The spermatogenesis of Amphiuma. Journ. Morphol., Vol. 15, 1899, Suppl., p. 57.

1875 Malbranc, M., Bemerkung über die Sinnesorgane der Seitenlinie bei Amphibien. Centralbl. med. Wiss., 1875, No. 1, p. 5.

1876 Malbranc, M., Von der Seitenlinie und ihren Sinnesorganen bei Amphibien. Zeitschr. wiss. Zool., Bd. 26, 1876, p. 24—86.

1893 Mall, Franklin P., Histogenesis of the retina in Amblystoma and Necturus. Journ. Morphol., Vol. 8, 1893, p. 415—432, 12 fig.

1839 Mandl, L., Dimensions des globules du sang chez le Proteus. Compt. Rend. Acad. Sc. Paris, T. 9, 1839, p. 739.

1839a Mandl, L., Note sur les globules sanguins du Protée et des Crocodiliens. Ann. Sci. nat., T. 12, 1839, p. 289—291, fig.

1875 Marchhetti, Di alcune nuove località del Proteus anguineus Laur. Boll. Soc. Adriat. Sc. nat., Vol. 1, 1875, p. 192—193.

1906 Maschkowski, Karl, Zur Entstehung der Gefäßendothelien und des Blutes bei Amphibien. Jena. Zeitschr., Bd. 41, 1906, p. 19—112, 6 Taf., 17 Fig.

1878 Marenzeller, Emil von, Ueber die drei lebenden japanischen Riesensalamander des k. k. Zoologischen Hofkabinetes. Verh. Zool.-bot. Ges. Wien, Bd. 27, 1878, p. 43.

1890 Marshall, Wm. B., Necturus maculatus in the Hudson River. Amer. Nat., Vol. 20, 1890, p. 779—780.

1888 Maurer, F., Die Kiemen und ihre Gefäße bei anuren und urodelen Amphibien. Morph. Jahrb., Bd. 13, 1888, p. 383—384.

1888a Maurer, F., Schilddrüse, Thymus und Kiemenreste bei Amphibien. Morph. Jahrb., Bd. 13, 1888, p. 296—382, 3 Taf., 6 Fig.

1906 Maurer, F., Die Entwicklung des Muskelsystems und der elektrischen Organe. Handbuch der Entwicklungslehre der Wirbeltiere von Oskar Hertwig, Bd. 3, Jena 1906.

1899 Mathews, Albert P., The changes in structure of the pancreas cell. Journ. Morphol., Suppl. to Vol. 15, 1899, p. 172—203, 3 pl.

1835 Mayer, A. F. J. C., Analecten für vergleichende Anatomie. Bonn 1835. 4°. p. 6 + 93 + 8, 7 Taf.

1817 Meckel, J. F., Ueber den Darmkanal der Reptilien. Deutsch. Arch. f. Physiol., Bd. 3, 1817, p. 199—232; Nachtrag, Bd. 5, 1819, p. 313—348.

1880 Merkel, Fr., Ueber die Endigungen der sensiblen Nerven in der Haut der Wirbeltiere. Rostock 1880.

1820 Merrem, Blas, Versuch eines Systems der Amphibien. Tentamen Systematis Amphibiorum. [German and Latin.] Marburg 1820. 8°. 1 Taf.

1870 MESTENHAUSER, J., Einige Betrachtungen über den Olm. Zool. Garten, Bd. 11, 1870, p. 365—368.

1893 MICHON, PIERRE DE, Sur la grande Salamandre du Japon. Bull. Soc. Sci. nat. Neuchâtel, T. 21, 1893, p. 186—187.

1875 MEYER, FR., Beitrag zur Anatomie des Urogenitalsystems der Selachier und Amphibien. Sitz.-Ber. Naturf. Ges. Leipzig, 1875, p. 38—44.

1889 MEYERHOFER, FRANZ, Untersuchungen über die Morphologie und Entwicklungsgeschichte des Rippensystems der urodelen Amphibien. Arb. Zool. Inst. Univ. Wien, 1909, p. 305—352, 2 Taf. u. 9 Fig.

1829 MICHAHELLES, C., Protens anguineus Aristoteli pronus ignotus fuit. Isis, 1829, p. 1270—1278.

1831 MICHAHELLES, C., Beiträge zur Naturgeschichte des Proteus. Isis, 1831, p. 499—508.

1898 MIHALKOVICS, V. von, Nasenhöhle und Jacobsonsches Organ. Eine morphologische Studie. Anat. Hefte, Bd. 11, 1898, p. 1—107, 11 Taf.

1862 MILNE, J., Bemerkungen über den Olm. Verh. d. Zool.-bot. Ges. Wien, Bd. 12, 1862, p. 87—88.

1900 MILLER, W. S., The brain of Necturus maculatus. Bull. Univ. Wisconsin, 1900, No. 33, p. 227—234, 2 pl.

1900a MILLER, W. S., The vascular system of Necturus maculatus. Bull. Univ. Wisconsin, Vol. 2, 1900, No. 33, p. 214—226, 2 pl.

1900b MILLER, W. S., The lung of Necturus maculatus. Bull. Univ. Wisconsin, Vol. 2, 1900, No. 33, p. 203—210, 6 pl.

1902 MILLER, W. S., The lymphatics of the lung of Necturus. Amer. Journ. Anat., Vol. 2, 1902, p. VI.

1905 MILLER, W. S., The blood and lymph vessels of the lung of Necturus maculatus. Amer. Journ. Anat., Vol. 4, 1905, p. 445—453, 3 fig.

1906 MILLER, W. S., The mesentery in Amphibia and Reptilia. Amer. Journ. Anat., Vol. 4, 1906, p. XIV—XV.

1874 MILNER, JAMES W., Report on the Fisheries of the Great Lakes. In United States Commission of Fish and Fisheries. Part 2. Report of the Commissioner for 1872 and 1873. Washington 1874. 8°. p. 78.

1892 MINOT, C. S., Human embryology. Philadelphia 1892. 8°.

1898 MINOT, C. S., On the veins of the Wolffian body in the Pig. Proc. Boston Soc. Nat. Hist., Vol. 28, 1898, p. 265—271, 1 fig., 1 pl.

1900 MINOT, C. S., On a hitherto unrecognized form of blood circulation without capillaries in the organs of Vertebrata. Proc. Boston Soc. Nat. Hist., Vol. 29, 1900, p. 185—215, 12 fig.

1822 MITCHILL, SAMUEL L., The Proteus of the North American Lakes. Amer. Journ. Sci. and Arts, Ser. 1, Vol. 4, 1822, p. 181—183.

1824 MITCHILL, SAMUEL L., Ueber einige, wahrscheinlich zum Geschlecht der Proteus gehörige Reptilien Neodamerikas. Frorieps Not., Bd. 7, 1824, p. 165—166.

1824a MITCHILL, SAMUEL L., Observations on several Reptiles of North America which seem to belong to the family Proteus. Amer. Journ. Sci. and Arts, Ser. 1, Vol. 7, 1824, p. 63—69, 2 pl.

1869 MIVART, ST. GEORGE, On the myology of Menopoma alleghaniensis, Menobranchus lateralis and Chamaeleon Parsonii. Proc. Zool. Soc. London, 1869 and 1870.

1869a MIVART, ST. GEORGE, Notes on the myology of Menobranchus lateralis. Proc. Zool. Soc. London, 1869.

1870 MIVART, ST. GEORGE, On the axial skeleton of the Urodela. Proc. Zool. Soc. London, 1870.

1887 MOBIUS, C., Der japanische Riesensalamander (Cryptobranchus japonicus) und der fossile Salamander von Oeningen (Andrias scheuchzeri). Neujahrsblatt Naturf. Ges. Zürich. 1887. 4°. p. 12.

1889 MONTGOMERY, HENRY, Some observations on the Menobranchus maculatus. Canadian Naturalist, Vol. 9, 1889, p. 159—164.

1878 MOSSO, ROSA, Su la morfologia comparata dei condotti escretori delle glandole gastriche nei Vertebrati. Boll. Sc. Pavia, 1878, Anno 2d, p. 33—39, 63—75, 101—108, 2 tab.

1879 MOSSO, ROSA, Ricerche anatomo-comparative sulla minuta inservazione degli organi trofici nei Cranioti inferiori. Torino 1879. 147 pp., 12 pl.

1908 MOODIE, ROY L., The ancestry of the caudate Amphibia. Amer. Nat., Vol. 42, 1908, p. 361—373.

1897 MORGAN, T. H., The development of the Frog's egg. New York 1897. 8°.

1903 MORGAN, T. H., Regeneration of the leg of Amphiuma means. Biol. Bull., Vol. 5, 1903, p. 203—216.

1832 MÜLLER, JOH. CHRISTOF, Beiträge zur Anatomie und Naturgeschichte der Amphibien. Zeitschr. Physiol., Bd. 4, 1832.

1864 MÜLLER, KARL, Der Riesensalamander. Natur, Bd. 13, 1864, p. 172—174.

1897 MURRAY, J. A., The vertebral column of certain primitive Urodela. Anat. Anz., Bd. 13, 1897, p. 651—664, 3 Fig.

1891 NAKAGAWA, ISAAC, The origin of the cerebral cortex and homologies of the optic lobe layers in the lower vertebrates. Journ. Morphol., Vol. 5, 1891, p. 1—10, 1 pl.

1828 NEILL, PATRICK, Some account of the habits of a specimen of Siren lacertina kept alive at Canonmills. Edinb. New Phil. Journ., Vol. 4, 1828, p. 336—355.

1832 NEILL, PATRICK, Siren. Isis, 1832, p. 698—699.

1902 NESSLER, A., Zur Frage über die Nerven des Darmkanales bei den Amphibien. Schriften Kais. Naturf. Ges. St. Petersburg, Bd. 32, 1902, p. 53—59.

NEUMAYER, L., Histogenese und Morphogenese des peripheren Nervensystems der Spinalganglien und des Nervus sympathicus. Handbuch der Entwicklungslehre der Wirbeltiere von OSKAR HERTWIG., Bd. 2, Jena 1906, p. 573—621.

NICOLAS, PAULETTE, Ueber die Hautdrüsen der Amphibien. Zeitschr. wiss. Zool., Bd. 56, 1893, p. 409—487, 3 Taf.

NUSSE, P., Proteus anguineus in der Gefangenschaft. Zool. Garten, Bd. 5, 1864, p. 241.

NORRIS, H. W., The development of the auditory vesicle in Necturus. Proc. Iowa Acad. Sci., Vol. 1, 1894, p. 105—107.

NORRIS, H. W., The so-called Dorso-trachealis branch of the seventh cranial nerve in Amphiuma. Anat. Anz., Bd. 27, 1905, p. 271—272.

NORRIS, H. W., The cranial nerve components in Amphiuma. Science, N. S. Vol. 27, 1908.

NORRIS, H. W., The cranial nerves of Amphiuma means. Journ. Comp. Neurol. and Psychol., Vol. 18, 1908, p. 527—568, 5 pl.

NUSBAUM, J., Ein Fall von Viviparität beim Proteus anguineus. Biol. Centralbl., Bd. 27, 1907, p. 370—375.

NUSBAUM, J., Zur Entwicklungsgeschichte der Occipitalregion des Schädels und der Wirbelsäule bei den Cyprinoiden. Krakau (Acad.) 1908. 8°.

OSTERTORN, ABRAHAM, Siren lacertina. Amoenitates Academicae, Vol. 7, 1789, p. 322.

OKAJIMA, KENJI, Zur Anatomie des inneren Gehirnganges von Cryptobranchus japonicus. Anat. Hefte, 154, 32, 1906, p. 233.

OKAJIMA, KENJI, Zur Anatomie des Geruchsorgans von Cryptobranchus japonicus. Anat. Anz., Bd. 29, 1906, p. 641—650, 5 Fig.

OKEN, LORENZ, Ueber den Olm Proteus anguineus. Isis, 1847, p. 641—645, 1 Fig.

OKEN, LORENZ, Allgemeine Naturgeschichte, Bd. 6, Stuttgart 1836, p. 417. 8°.

OPPEL, A., Beiträge zur Anatomie des Proteus anguineus. Arch. mikr. Anat. u. Entw., Bd. 34, 1890.

OPPEL, A., Die Magendrüsen der Wirbeltiere. Anat. Anz., Bd. 11, 1896, p. 506—602.

OPPEL, A., Lehrbuch der vergleichenden mikroskopischen Anatomie der Wirbeltiere. Jena 1896.

OSAWA, G., Beiträge zur Anatomie des japanischen Riesensalamanders. Festschrift TAGUCHI, Tokio 1899.

OSAWA, G., Beiträge zur Anatomie des japanischen Riesensalamanders. Mitt. Med. Facult. Tokio, Vol. 5, 1902, p. 8—40, 7 Taf.

OSBORN, H. F., Preliminary observations upon the brain of Amphiuma. Proc. Acad. Nat. Sci. Philadelphia, 1883, p. 177—184.

OSBORN, H. F., Preliminary notes upon the brain of Menopoma. Proc. Acad. Nat. Sci. Philadelphia, 1884.

OSBORN, H. F., Preparing brain of Urodela. Journ. Roy. Micr. Soc., (2) Vol. 5, 1885, p. 534—537.

OSBORN, H. F., The origin of the corpus callosum; a contribution upon the cerebral commissures of the vertebrata. Morphol. Jahrb., Bd. 12, 1886, p. 223—251.

OSBORN, H. F., A contribution to the internal structure of the Amphibian brain. Journ. Morph., Vol. 2, 1888, p. 51—92, 3 pl.

OSBORN, J., Observations upon the urodele Amphibian brain. Zool. Anz., Bd. 7, 1884, p. 679—682.

OWEN, RICHARD, On the structure of the heart in the Perennibranchiata. Trans. Zool. Soc. London, 1835.

OWEN, RICHARD, On the blood-disks in Siren lacertina. Micr. Journ., 1842.

OWEN, RICHARD, On the anatomy of Vertebrates. Vol. 3. London 1866. 8°.

PARKER, G. H., Variations in the vertebral column of Necturus. Anat. Anz., Bd. 11, 1896, p. 711—717, 2 Fig.

PARKER, W. K., A monograph on the structure and development of the shoulder girdle and sternum in the Vertebrata. Ray Soc. London, 1868.

PARKER, W. K., On the structure and development of the skull in the Urodelous Amphibia. Phil. Trans. Roy. Soc. London, Vol. 167, 1877, p. 529—597, 9 pl.

PETER, KARL, Ueber die Bedeutung des Atlas der Amphibien. Anat. Anz., Bd. 10, 1894, p. 565—571.

PETER, KARL, Die Entwicklung und funktionelle Gestaltung des Schädels von Ichthyophis glutinosus. Morphol. Jahrb., Bd. 25, 1898, p. 555—628, 3 Taf., 1 Fig.

PUGNAT, C., Action physiologique du venin de Salamandre du Japon (Sieboldia maxima). Antivention par la chaleur et vaccination de la grenouille contre ce venin. Compt. Rend. Soc. Biol. Paris, T. 4, 1897, p. 723—725.

PUGNAT, C., Propriétés immunisantes du venin de Salamandre du Japon vis-à-vis du venin de vipère. Compt. Rend. Soc. Biol. Paris, T. 4, 1897, p. 822—823.

PIERSOL, G. A., Structure of spermatozoa (Amphiuma tridactylum). Univ. Med. Magaz. Philad, 1889, p. 20.

PLATT, JULIA B., Ontogenetische Differenzierung des Ektoderms in Necturus. 1. Studie. Arch. f. mikr. Anat. u. Entw., Bd. 43, 1894, p. 911—966, 6 Taf.

PLATT, JULIA B., The development of the thyroid gland and of the suprapericardial bodies in Necturus. Anat. Anz., Bd. 11, 1896, p. 557—567, 9 Fig.

1896a PLATE, JULIA B., Ontogenetic differentiations of the ecoderm in Necturus. Study 2. On the development of the peripheral nervous system. Quart. Journ. Micr. Sci., Vol. 38, 1896, p. 485—547, 2 pl.

1897 PLATE, JULIA B., Development of the cartilaginous skull and of the branchial and hypoglossal musculature in Necturus. Morphol. Jahrb., Bd. 35, 1897, p. 377—464, 3 pl.

1859 POOCK VAN MEERDERVOORT, J. L. C., Over den grooten Japanschen Salamander. Natuurk. Tijdschr. Nederl. Indië, Ser. 4, Vol. 6, 1859, p. 386.

1904 PRENANT, A., Sur la morphologie des cellules épithéliales ciliées qui recouvrent le péritoine hepatique des Amphibiens. Compt. Rend. Soc. Biol. Paris, T. 55, 1903, p. 1041—1046.

1895 RABL, H., Ueber die Herkunft des Pigments in der Haut der Larven der urodelen Amphibien. Anat. Anz., Bd. 10, 1895, p. 12—16.

1889 RABL, CARL, Theorie des Mesoderms. Morphol. Jahrb., Bd. 15, 1889, p. 113—252, 4 Taf., 9 Fig.

1898 RABL, CARL, Ueber den Bau und die Entwicklung der Linse. 1. Selachier und Amphibien. Zeitschr. wiss. Zool., Bd. 63, 1898, p. 496—572, 4 Taf., 14 Fig.

1819 RAFINESQUE, Prodrome de soixante-dix nouveaux genres d'animaux découverts dans l'intérieur des Etats Unis d'Amérique durant l'année 1818. Journ. de Phys., T. 88, 1819, p. 417—429.

1825 RATHKE, H., Ueber die Entwicklung der Geschlechtsteile bei den Amphibien. Beiträge zur Geschichte der Tierwelt. 3 Abt. Halle 1825.

1894 REESE, A. M., The sexual elements of the Giant Salamander. Amer. Nat., Vol. 38, 1894, p. 497.

1894a REESE, A. M., The sexual elements of the Giant Salamander, Cryptobranchus alleghaniensis. Biol. Bull., Vol. 6, 1904, p. 220—223.

1894b REESE, A. M., The enteron and integument of Cryptobranchus alleghaniensis. Trans. Amer. Micr. Soc., Vol. 24, 1894, p. 109—130.

1906 REESE, A. M., Integument von Cryptobranchus. Trans. Amer. Micr. Soc., Vol. 26, 1906, p. 109—130, 2 pl.

1906a REESE, A. M., The eye of Cryptobranchus. Biol. Bull., Vol. 9, 1906, p. 22—26, 4 fig.

1905 REESE, A. M., Anatomy of Cryptobranchus alleghaniensis. Amer. Nat., Vol. 40, 1906, p. 287—326.

1906a REESE, A. M., Observations on the reactions of Cryptobranchus and Necturus to light and heat. Biol. Bull., Vol. 11, 1906, p. 93—99.

1876 REIN, J. J., and ROETZ, A. von, Beitrag zur Kenntnis des Riesensalamanders (Cryptobranchus japonicus). Zool. Garten, Bd. 17, 1876, p. 33—37.

1884 REIN, J. J., Reptiles and batrachians (Japan). New York 1884. p. 480.

1897 REZÁČEK, J., Histologie sítě od Cryptobranchus japonicus. Rozpr. Česko. Akad., Ser. 2, 1897, p. 10.

1897a REZÁČEK, J., L'histologie de l'oeil de Cryptobranchus japonicus. Bibliogr. Anat., Année 5, 1897, p. 139—146.

1884 RETZIUS, G., Das Gehörorgan der Wirbeltiere. 1. Das Gehörorgan der Fische und Amphibien. Stockholm 1884.

1901 RETZIUS, G., Zur Kenntnis der Limitans externa der nervösen Zentralorgane. Biol. Unters. RETZIUS, Bd. 11, 1901, p. 77—84.

1905 RETZIUS, G., Zur Kenntnis der Nervenendigungen in den Papillen der Zunge der Amphibien. Biol. Unters. RETZIUS, Bd. 12, 1905, p. 61—64.

1897 REYNOLDS, SIDNEY H., The vertebrate skeleton. Cambridge 1897. 8°.

1907 RIBBING, L., Die distale Armmuskulatur der Amphibien, Reptilien und Säugetiere. Zool. Jahrb., Bd. 23, 1907, p. 587—683, 2 Taf.

1902 ROGUSKI, ALEX V., Ueber die Struktur und die Bedeutung der LANGERHANS-schen Inseln im Pancreas der Amphibien. Berlin 1902. p. 37.

1909 KINGEL, OSCAR, The rate of digestion in cold-blooded vertebrates. The influence of season and temperature. Amer. Journ. Physiol., Vol. 24, 1909, p. 447—458.

1905 ROMPELDT, THEODOR H., A case of abnormal venous system in Necturus maculatus. Amer. Nat., Vol. 39, 1905, p. 391—396.

1907 RUST, PETRONELLA JOHANNA nr., Die Entwicklung des Herzens, des Blutes und der großen Gefäße bei Megalobatrachus maximus SCHLEGEL. Jena. Zeitschr., Bd. 35, 1907, p. 309—346, 6 Taf.

1901 RUSCONI, W. J., Ueber den Bau des Gehirns der Amphibien. XI. Vers. russ. Naturf. u. Aerzte, St. Petersburg 1901.

1903 ROGACHEWA, W. J., Zur Morphologie des Gehirns der Amphibien. Arch. mikr. Anat. u. Entw., Bd. 62, 1903, p. 207—213, 2 Taf.

1817 RUSCONI, M. A., Ueber den Proteus anguineus. Isis, 1817, p. 1017—1019.

1897 RULE, GEORG, Ueber das peripherische Gebiet des Nervus facialis bei Wirbeltieren. Festschrift GEGENBAUR, Bd. 3, 1897, p. 199—348, 70 Fig.

1818 RUSCONI, M., and CONFIGLIACHI, Del Proteo anguineo di LAURENTI monografia. Pavia 1819.

1827 RUSCONI, MAURO, Descrizione di un Proteo femmina notabile per lo sviluppo delle parti della generazione. Isis, 1827, p. 74.

1828 RUSCONI, MAURO, Sopra un Proteo femmina. Pavia 1828. 4°. 1 tab.

1837 Rusconi, Mauro, Observations anatomiques sur la Siréne mise en parallèle avec le Protée et le Lézard de la Salamandre aquatique. Pavie 1837. 4°. 6 pl.

1843 Rusconi, Mauro, Osservazioni sopra il Proteo. Milano 1843. 8°

1843a Rusconi, Mauro, Nuove osservazioni sopra il Proteo anguino. di. Lettera. Gazz. dell'Istit. Lomb. e Ibid. ital., Vol. 6, 1843, p. 288—293.

1879 Ryder, John A., Morphological notes on the limbs of the Amphiumidae, as indicating a possible synonymy of the supposed genera. Proc. Acad. Sci. Philad., 1879, p. 11—15.

1880 Ryder, John A., On a brood of larval Amphiuma. Amer. Nat., Vol. 23, 1880, p. 827—828.

1883 Saint-Hilaire, Const., Bau des Darmepithels bei Amphiuma. Anat. Anz., Bd. 22, 1903, p. 483—484, 6 Fig.

1887 Sasaki, C., Some notes on the Giant Salamander of Japan. Journ. Coll. of Sci. Tokyo, Japan, Vol. 1, 1887, p. 269—271.

1902 Schäfer, A., Ueber kontraktile Fibrillen in den glatten Muskelfasern des Mesenteriums der Urodelen. Anat. Anz., Bd. 22, 1902, p. 65—82, 2 Taf., 6 Fig.

1906 Schauinsland, H., Die Entwicklung der Wirbelsäule, nebst Rippen und Brustbein. Handbuch der Entwicklungslehre der Wirbeltiere von Oskar Hertwig, Bd. 3, 1906, p. 339—572.

1891 Schlampp, K. W., Die Augenlinse des Proteus anguineus. Biol. Centralbl., Bd. 11, 1891, p. 101—112.

1892 Schlampp, K. W., Das Auge des Grottenolmes (Proteus anguineus). Zeitschr. wiss. Zool., Bd. 53, 1892, p. 537—557, 1 Taf.

1837 Schreibers, H., Abbildungen neuer oder unvollständig bekannter Amphibien nach der Natur oder dem Leben entworfen. Düsseldorf 1837—1844. kl. Fol. 50 Taf.

1838 Schlegel, H., Fauna Japonica. Lugd. Bat. 1838, p. 127, 2 tab.

1850 Schmidt, A., Notizen über die von ihm aus der Planina-Höhle mitgebrachten und der Klasse vorgezeigten Proteen. Sitz.-Ber. math.-nat. Akad. Wien. in Wien, Bd. 5, p. 229—232.

1864 Schlegel, F. J. J., Gudmand, Q. J., en van den Hoeven, J., Aanteekeningen over de anatomie van den Cryptobranchus japonicus. Natuurk. Verh. Maatsch. Wetensch. Haarlem, Vol. 19, 1864, p. 66, 12 pl.

1878 Schmidt, H. D., The structure of the coloured bloodcorpuscles of Amphiuma tridactylum, the frog and man. Journ. Roy. Mier. Soc., Vol. 1, 1878, p. 57—78 and 97—120.

1904 Schuster, Victor, Studien über Ovogenese. 1. Die Wachstumsperiode der Eier von Proteus anguineus. Anat. Hefte, Bd. 27, 1904, p. 1—74, 4 Taf.

1779 Schneider, J. G., Historiae Amphibiorum naturalea et litterariae. Fasciculus primus. Jenae 1779. 8°.

1873 Schreiber, O., Ueber den Olm. Gaea, Bd. 9, 1873, p. 524.

1890 Schneider, R., Neue histologische Untersuchungen über die Eisenaufnahme in den Körper des Proteus. Sitz.-Ber. K. Preuß. Akad. Wiss. Berlin, Bd. 36, 1890, p. 882—897.

1904 Strickershen, Walther, Die Brutpflege bei den Amphibien und besonders bei dem japanischen Riesensalamander (Megalobatrachus maximus). Prometheus, Bd. 16, 1904, p. 37—40, 52—54, 15 Fig.

1801 Schreibers, Ch., A historical and anatomical description of a doubtful amphibious animal of Germany called by Laurenti Proteus anguineus. Phil. Trans. Roy. Soc. London, Vol. 91, 1801, p. 24, 2 pl.

1818 Schreibers, Ch., Proteus anguineus. Vienne 1818. 4°.

1820 Schreibers, Ch., Sur le Protée. Isis, 1820, p. 567—570.

1893 Schreiner, A., Beiträge zur Kenntnis der Amphibienhaut. Zool. Jahrb., Bd. 6, 1893, p. 481—490, 1 Taf.

1900 Schultz, Paul, Ueber die Anordnung der Muskulatur im Magen der Batrachier. Arch. Anat. Physiol., Phys. Abt., 1900, p. 1—8, 4 Fig.

1861 Schulze, F. E., Ueber die Nervenendigung in den sogenannten Schleimkanälen der Fische und über entsprechende Organe der durch Kiemen atmenden Amphibien. Arch. Anat. Physiol., 1861, p. 759—769, 1 Taf.

1870 Schulze, F. E., Ueber die Sinnesorgane der Seitenlinie bei Fischen und Amphibien. Arch. mikr. Anat. u. Entw., Bd. 6, 1870, p. 62, 3 Taf.

1876 Schulze, F. E., Zur Fortpflanzungsgeschichte des Proteus anguineus. Zeitschr. wiss. Zool., Bd. 26, 1876, p. 350—354.

1905 Schultze, O., Ueber partiell albinotische und mikrophthalmische Larven von Salamandra maculata nebst einigen Angaben über die Fortpflanzung dieses Tieres. Zeitschr. wiss. Zool., Bd. 82, 1905, p. 472—493.

1905 Seydel, O., Ueber die Nasenhöhle und das Jacobsonsche Organ der Amphibien. Morphol. Jahrb., Bd. 23, 1895, p. 453—543, 22 Fig.

1828 Siebold, Carol. Th. Ernst, Observationes quaedam de salamandris et tritonibus. Berolini 1828, p. 6 + 30, 1 tab.

1833 Siebold, Philipp Franz von, Fauna japonica. Leyden 1833—1851.

1904 Sierlinger, F., Zur Anatomie der Urodelenextremität. Zool. f. Anat. u. Entw., 1904, p. 385—404, 1 Taf.

1906 Smallwood, W. M., The sacrum of Nectarus. Anat. Anz., Bd. 33, 1906, p. 237—239, 1 Fig.

1890 Smirnow, A., Die Struktur der Nervenzellen im Sympathicus der Amphibien. Arch. mikr. Anat. u. Entw., Bd. 35, 1890, p. 407—424, 2 Taf.

1900 Smith, F., Some additional data on the position of the sacrum in Nectarus. Amer. Nat., Vol. 34, 1900, p. 635—638.

1906   Smith, Bertram, G., Preliminary report on the embryology of Cryptobranchus alleghaniensis. Biol. Bull., Vol. 9, 1906, p. 146—164, 1 pl.

1898   Smith, Hugh M., On the occurrence of Amphiuma, the so-called Congo Snake in Virginia. Proc. U. S. Nat. Mus., Vol. 21, 1898, p. 370—380.

1882   Smith, J., Leix, 1882, p. 1688.

1828   Smith, F. Aug., Account of the dissection of a Proteus of the Lakes (Menobranchus) with remarks on the Siren Intermedia. Ann. Lyc. Nat. Hist. New York, Vol. 2, 1828, p. 250—263.

1897   Schotte, J., Die Forchung des Wirbeltiereins. Ergeb. Anat. Entw., Bd. 6, 1897, p. 493—503, 38 Fig.

1892   Schotte, J., Ueber die Entwicklung des Blutes, des Herzens und der großen Gefässstamme der Salmoniden, nebst Mitteilungen über die Ausbildung der Herzform. Anat. Hefte, 1892, Heft 63, p. 579—698.

1869   Stokes, W. W., Proteus anguineus. Sci. Gossip, 1869, p. 135.

1892   Stejneger, Leonhard, Preliminary description of a new genus and species of blind cave Salamander from North America (Typhlotriton spelaeus. Proc. U. S. Nat. Mus., Vol. 15, 1892, p. 115—117, 1 pl.

1896   Stejneger, Leonhard, Description of a new genus and species of blind cave Batrachians from the subterranean waters of Texas. Proc. U. S. Nat. Mus., Vol. 18, 1896, p. 619—621.

1907   Stejneger, Leonhard, Herpetology of Japan and adjacent territory. Washington 1907. 8°.

1907a   Stejneger, Leonhard, Specific name of Necturus maculosus. Science, Vol. 25, 1907, p. 130.

1906   Stohr, Philipp, Histology arranged upon an embryological basis by Frederic T. Lewis. Philadelphia 1906.

1870   Strauch, Alexander, Revision der Salamandriden-Gattungen nebst Beschreibung einiger neuen oder weniger bekannten Arten dieser Familie. Mem. Ac. Sc. St. Pétersbourg, T. 16, 1870, p. 100, 2 pl.

1872   Stricker, S. (ed.), Manual of human and comparative histology. Tr. by Henry Power. 3 vols. London 1872.

1892   Strong, Oliver S., The structure and homologies of the cranial nerves of the Amphibia as determined by their peripheral distribution and internal origin. Anat. Anz., Bd. 7, 1892, p. 467—471.

1895   Strong, Oliver S., The cranial nerves of Amphibia. Journ. Morphol., Vol. 10, 1895, p. 101—230, 6 pl.

1888   Stahl, K. W., [Ueber den japanischen Salamander]. [Russian.] Nachr. d. Kais. Ges. d. Freunde d. Naturwiss. Moskau, Bd. 56, 1888.

1874   Stieda, Tu., Ueber die Epidermis der Amphibien didactyla [menu]. Mitt. Naturf. Ges. Bern, 1874, p. 48.

1897   Studnicka, F. K., Studien über den Bau des Sehnerven der Wirbeltiere. Jen. Zeitschr., Bd. 31, 1897, p. 1—28, 2 Taf.

1909   Takahashi, Kataro, Histogenesis of the lateral line system in Necturus. Thesis, Chicago 1909.

1838   Temminck, C. J., and Schlegel, H., Fauna Japonica auctore Pn. Fr. de Siebold. Reptilia elaborantibus C. J. Temminck et H. Schlegel. Cum mappa geographico-zoologica et tabellis lithogr. XXXVIII. Lugduni Batavorum, ex officina lithogr. auctoris et typis J. G. Lalau, 1838. Fol. XXII + 144 pp. + doublepage map + 2 + 10 + 8 pl.

1904   Terry, Robert J., Two skulls of larval Necturus. Amer. Journ. Anat., Vol. 3, 1904, p. XI.

1906   Terry, Robert J., The nasal skeleton of Amblystoma punctatum (Lin.). Trans. St. Louis Acad. Sci., Vol. 16, 1906, p. 95—124, 4 pl.

1875   Tocagnani, M., Sulla diffusione del Proteus. Bollet. Soc. Adriat. Sc. nat., Vol. 1, 1875, p. 152—156.

1882   Tornier, Chas. H., Habits of Menopoma. Amer. Nat., Vol. 16, 1882, p. 139—140.

1820   Treviranus, De protei anguinei encephalo et organis sensuum disquisitiones zootomicae. Commentationes societatis regiae scientiarum Gottingensis recentiores, Göttingen 1820, Vol. 4.

1837   Tschudi, J. J., Ueber den Ilison diluvii testis (Andrias Scheuchzeri). Neues Jahrbuch für Mineralogie, Geognosie, Geologie und Petrefaktenkunde, Stuttgart 1837, Heft 5, p. 545—547.

1838   Tschudi, J. J., Classification der Batrachier mit Berücksichtigung der fossilen Tiere dieser Abteilung der Reptilien. Mém. Soc. Sci. nat. de Neuchatel, Vol. 2, 1838.

1903   Tuckerman, W., Ueber das Rückenmark von Cryptobranchus japonicus. Leipzig 1903. p. 21, 9 Taf.

1862   Vaillant, Léon, Note sur la structure du noyau des globules sanguins, et la composition de l'encéphale chez la Sirène lacertina. Compt. Rend. et Mém. Soc. d. Biol. Paris, Sér. 3, T. 4, 1862, p. 4—6.

1863   Vaillant, Léon, Mémoire pour servir à l'histoire anatomique de la sirène lacertine. Ann. Sci. nat. Zool, Sér. 4, T. 19, 1863, p. 295—346, 3 pl.

1863a   Vaillant, Léon, Note sur l'anatomie de la sirène lacertine. Compt. Rend. Acad. Sc, Paris, T. 56, 1863, p. 830.

1837   Valentin, G., Bruchstücke aus der feineren Anatomie des Proteus anguineus. Report. f. Anat. u. Physiol, Bd. 1, 1837, p. 292—298.

1841   Valentin, G., Ueber die Samenfadenbündel und die Afterdrüse des Proteus anguineus. Report. f. Anat. u. Physiol, Bd. 6, 1841, p. 358—358.

1903   Van Pee, P., Recherches sur le développement des extrémités chez Amphiuma et Necturus. Compt. Rend. Assoc. Anat. Sess. 5. Liège, 1903, p. 37—42.

1903a   Van Pee, P., Ueber die Entwicklung der Extremitäten bei Amphiuma und Necturus. Verh. Anat. Ges. Heidelberg, 1903, p. 83—86.

1894 Van Pee, P., Les membres chez Amphiuma. Anat. Anz., Bd. 24, 1904, p. 176—182, 4 Fig.

1857 Vaayon (?), On the Proteus anguineus. Unsvorsworth's Magaz. Nat. Hist., Vol. 1, 1857, p. 625—632.

1894 Virchow, Hans, Einige embryologische und angiologische Catalogunges über neolasserikanische Wirbeltiere. Sitz.-Ber. Ges. naturf. Freunde Berlin, 1894, p. 33—44.

1904 Voss, A., Sur quelques expériences effectuées au laboratoire des Catacombes du Muséum d'Histoire naturelle. Compt. Rend. Acad. Sci. Paris, T. 138, 1904, p. 706—708.

1828 Wagler, Jon., Descriptiones et icones Amphibiorum. Stuttgartiae 1828—1833. fol. 36 tab.

1830 Wagler, Jon., Natürliches System der Amphibien mit vorgehender Classification der Säugetiere und Vögel. München 1830. 8°.

1889 Wagner, Franz, Der Olm (Proteus anguineus). Isis Dresd., Bd. 5, 1889, p. 414—417.

1837 Wagner, Rud., Notes on Proteus anguineus. Proc. Zool. Soc. London, Vol. 5, 1837, p. 107—108.

1897 Waite, F. C., Variations in the brachial and lumbo-sacral plexi of Necturus maculosus Rafinesque. Bull. Mus. Comp. Zool. Harvard Univ., Vol. 31, 1897, p. 69—92.

1907 Waite, F. C., Specific name of Necturus maculosus. Amer. Nat., Vol. 41, 1907, p. 23—30.

1887 Walter, Ferdinand, Das Visceralskelet und seine Muskulatur bei den einheimischen Amphibien und Reptilien. Jen. Zeitschr., Bd. 21, 1887, p. 1—45, 4 Taf.

1905 Warren, John, The development of the paraphysis and the pineal region in Necturus maculatus. Amer. Journ. Anat., Vol. 5, 1905, p. 1—27, 23 fig.

1881 Wedzivel, V., Zur Kenntnis des weiblichen Proteus anguineus. Sitz.-Ber. K. böhm. Ges. Prag., 1881, p. 297—303, 1 Taf.

1878 Weismann, A., and Weissbachim, R., Zwei noch junge Siren lacertina. Aus dem zoologischen und anatomischen Institut der Universität Freiburg i. Br. Zool. Anz., Bd. 1, 1878, p. 6.

1890 Wellkir, W. N., Ueber Mehrzahl der Lymphherzen bei Proteus anguineus und Rana temporaria. Trav. Soc. Natur. St. Pétersbourg, Sect. Zool., Vol. 20, 1890.

1906 Whipple, Inez, The epistloid apparatus of Urodeles. Biol. Bull., Vol. 10, 1906, p. 255—297.

1885 Whitman, C. O., Methods of research in microscopical anatomy and embryology. Boston 1885. 8°. p. 255.

1888 Whitman, C. O., The eggs of Amphibia. Amer. Nat., Vol. 22, 1888, p. 867.

1888a Whitman, C. O., Some new facts about the Hirudinea. Journ. Morphol., Vol. 2, 1888, p. 586—590.

1899 Whitman, C. O., Animal behavior. Biological Lectures from the Marine Biological Laboratory Woods Holl, Mass., 1898, Boston. 8°. p. 285—338.

1896 Whitman, C. O., and Eycleshymer, A. C., The egg of Amia and its cleavage. Journ. Morphol., Vol. 12, 1896, p. 309—356, 1 pl.

1865 Wied-Neuwied, Prinz Maximilian zu, Verzeichnis der Reptilien, welche auf einer Reise im nördlichen Amerika beobachtet wurden. Nov. Acta Acad. Leop.-Carol., Vol. 32, 1865.

1877 Wiedersheim, R., Zur Fortpflanzungsgeschichte des Proteus anguineus. Morphol. Jahrb., Bd. 3, 1877, p. 632.

1877a Wiedersheim, R., Das Kopfskelet der Urodelen. Morphol. Jahrb., Bd. 3, 1877, p. 352—448, 5 Taf., 1 Fig.; p. 459—548, 4 Taf., 5 Fig.

1877b Wiedersheim, R., Ueber Neubildung von Kiemen bei Siren lacertina. Morphol. Jahrb., Bd. 3, 1877, p. 630—632.

1890 Wiedersheim, R., Beiträge zur Entwicklungsgeschichte von Proteus anguineus. Arch. mikr. Anat. u. Entw., Bd. 35, 1890, p. 121—140, 2 Taf.

1892 Wiedersheim, R., Das Gliedmaßenskelett der Wirbeltiere mit besonderer Berücksichtigung des Schulter- und Beckengürtels bei Fischen, Amphibien und Reptilien. Mit 40 Figuren im Texte und einem Atlas von 17 Tafeln. Jena 1892. 8°. p. 267.

1909 Wiedersheim, R., Lehrbuch der vergleichenden Anatomie der Wirbeltiere. 1909.

1874 Wilder, B. G., Menobranchus edible. Amer. Nat., Vol. 8, 1874, p. 488.

1889 Wilder, B. G., On the habits of Cryptobranchus. Amer. Nat., Vol. 10, 1889, p. 816—817.

1891 Wilder, H. H., A contribution to the anatomy of Siren lacertina. Zool. Jahrb., Bd. 4, 1891, p. 653—696.

1892 Wilder, H. H., Die Nasengegend von Menopoma alleghaniense und Amphiuma tridactylum nebst Bemerkungen über die Morphologie des Ramus ophthalmicus profundus trigemini. Zool. Jahrb., Abt. f. Anat. u. Ontog., Bd. 6, 1892, p. 155—176, 2 Taf.

1896 Wilder, H. H., The amphibian larynx. Zool. Jahrb., Bd. 9, 1896, p. 273—318, 3 Taf., 4 Fig.

1903 Wilder, H. H., The skeletal system of Necturus maculatus Rafinesque. Mem. Boston Soc. Nat. Hist., Vol. 5, 1903, p. 387—439, 6 pl., 1 fig.

1909 Wilder, Inez W., The lateral nasal gland in Amphiuma. Journ. Morphol., Vol. 20, 1909, 1 pl., 7 fig.

1898 Winslow, G. M., The chondrocranium of the Ichthyopsida. Bull. Essex Inst., Vol. 28, 1898, p. 87—141, 1 pl.

1904 Winslow, G. M., Three cases of abnormality in Urodeles. Tufts Coll. Stud., 1904, No. 8, p. 387—414, 2 pl.

1854 Wyman, Jeffries, Structure of the heart and physiology of respiration in the Menobranchus and Batrachians. Proc. Boston Soc. Nat. Hist., Vol. 6, 1854, p. 51—52.

1888 ZELLER, E., Ueber die Larve des Proteus anguineus. Zool. Anz., Bd. 11, 1898, p. 570—572.

1889 ZELLER, E., Ueber die Fortpflanzung von Proteus anguineus und seine Larve. Jahreshefte d. Vereins f. vaterl. Naturk. in Württemberg, 1889.

1890 ZELLER, E., Ueber die Befruchtung der Urodelen. Zeitschr. wiss. Zool., Bd. 49, 1890, p. 583—601, 3 Fig.

1895 ZUCKERKANDL, E., Zur Anatomie und Entwicklungsgeschichte der Arterien des Unterschenkels und des Fußes. Anat. Hefte, Bd. 5, 1895, p. 207—291, 6 Taf.

1895a ZUCKERKANDL, E., Zur Anatomie und Entwicklungsgeschichte der Arterien des Vorderarmes. Anat. Hefte, Bd. 5, 1895, p. 157—205.

1897 ZWICK, WILH., Beiträge zur Kenntnis des Baues und der Entwicklung der Amphibiengliedmaßen, besonders von Carpus und Tarsus. Zeitschr. wiss. Zool., Bd. 63, 1897, p. 62—114, 4 Taf.

B. Alphabetical arrangement of authors under topics.

I. General Works.

a) Systematic.

Baird 49.
Barnes 26, 27.
Barton 67, 68, 12, 12a.
Beauvois 20.
Boulenger 82.
Camper 68.
Clapisson 79, 93.
Cope 85, 86, 89, 89a.
Cuvier 00, 27, 31.
Eckel and Paulmier 02.
Faber 64.
Fitzinger 26, 50.
Freyer 66.
Garnier 88.
Gibbes 54, 53.
Gray 25, 57, 73.
Hallowell 56, 58.
Harlan 27, 33, 67c.
Harting 71.
May 92.
Van der Hoeven 38b, 39, 66a, 66b.
Holbrook 36.
Kingsbury 94.
Kingsley 92.
Kneeland 56.
Lacépède 97, 07a.
Le Conte 26, 29.
Leuckart 49.
Merrem 20.
Rafinesque 19.
Rein 84.
Rein and von Resetz 76.
Schlegel 38.
Siebold 33.
Steineger 92, 96, 97, 97a.
Strauch 70.
Temminck and Schlegel 38.
Tschudi 37, 38.
Wagler 30.
Wasse 07.
Wiedersheim 66.

b) Zoological.

Abbott 78.
Andres 96.
Beale 78.
Beddard 03, 04.
Bell 35.
Bendes 07.
Berry 98.
Bettzieck 69.
Bjeletzkij 89.
Blanchard 71, 71a, 94.
Bois 26.
Bonaparte 32, 38, 39.
Bugnion 74.
Chiaje 40.
Claus 82.
Genfglischt and Rusconi 21.
Cope 96.
Dalton 50.
Daviess 94, 95, 96.
De Kay 42.
Dexler 59.
Dumeril et Bibron 34.
Ecker 59.
Ehrenberg 62, 67, 68, 70, 72, 74.
Eigenmann 99.
Kiessann 75.
Ellis 66.
Emerson 05.
Emery 97.
Erber 63, 77.
Eycleshymer 00.
Fischer, J. G. 64.
Fischer, Sigwart 00.
Frauenfeld 61.
Frear 82.
Fullebore 94.
Galow 01.
Geerts 88, 84.
Gegenbaur 78.
Geyer 05.
Gmelin 86.
Grato 70, 70a.
Grube 66.
Goerne 86.

Hargitt 92.
Harlan 25, 24, 24a, 26, 26a.
Hay 88.
Heidenhain 07.
Hertwig 06.
Van der Hoeven 38, 38a, 67, 67a, 67b.
Hoffmann 77.
Humboldt 05.
Hunter 66.
Huxley 74, 75.
Hyrtl 50, 65, 65a.
Ishikawa 00, 02, 04, 08.
James 23.
Kammerer 05, 05a.
Kingsley 92.
Kneeland 58.
Kölliker 89.
Korshuber 84.
Lankes 06.
Le Conte 49.
Leuckart 21.
Lewis 04.
Leydig 53.
Linnaeus 66.
Lloyd 66.
Marchesetti 75.
Marenzeller 78.
Marshall 93.
Mayer 35.
Pouspe van Meerdervoort 59.
Mettenheimer 70.
Meuron 93.
Michahelles 29, 31.
Milde 62.
Milner 74.
Minot 92.
Mitchill 22, 24, 24a.
Moesch 87.
Montgomery 90.
Moodie 08.
Morgan 03.
Müller, J. C. 32.
Müller, K. 64.
Neill 28, 32.
Nobbe 64.

Nussbaum 07.
Oesterlen 80.
Oken 17, 36.
Oppel 80, 96.
Osawa 96, 02.
Osborn 83.
Owen 66.
Phisalix 97, 97a.
Rabl 80.
Reese 06.
Regaud 97.
Rudolphi 17.
Rusconi 27, 28, 37, 43, 17a.
Ryder 88.
Rusconi and Configliachi 18.
Naski 87.
Schlegal 37.
Schmidt 60.
Schmidt, Goddard en Van der Hoeven 64.
Schneider, J. G. 79.
Schneider, O. 73.
Schneider, R. 90.
Schreibers 01, 18, 20.
Siebold 28.
Smith, H. M. 98.
Smith, J. 32.
Smith, F. A. 28.
Spicer 69.
Stöhr 06.
Stricker 72.
Strube 88.
Tommasini 73.
Townsend 82.
Treviranus 20.
Vaillant 63, 63a.
Valentin 37.
Visier 37.
Wagler 28.
Wagner, F. 90.
Wagner, R. 37.
Weismann 81.
Weismann and Wiedersheim 78.
Whitman 85, 99.
Wiedersheim 77, 90, 01.
Wilder, B. G. 74, 83.
Wilder, H. H. 91.
Zeller 89.

c) Embryological.

Broili 03.
Chauvin 83, 83a.
Cope 86.
Kyoleshymer 02, 04, 04a.
Fullebon 94.
May 88.
Kammerer 07.
Keibel 01, 06.

Kerbert 03, 04, 06.
Krahn 63.
Mc Gregor 96.
Morgan 97.
Platt 90, 96.
Rabl 98.
Schneumbru 04.
Schulze, F. E. 76.
Schulze, O. 03.
Smith, B. G. 06.
Virchow 94.
Yaró 03.
Whitman 88, 88a.
Wiedersheim 77, 90.
Zeller 89, 90.

2. Earliest Stages.

Andrews 88, 88a.
Bunn 04, 05.
Camerano 83.
Kyoleshymer 95, 02.
Ishikawa 03, 05.
Kerbert 03.
Lange 06, 07.
Schmidt 04.
Sobotta 97.
Whitman and Kyoleshymer 96.
Zeller 90.

3. Skeleton.

Albrecht 78.
Baur 85, 88, 89, 91.
Baetke 02.
Carlson 86.
Cope 89.
Davison 97.
Dugès 34.
Field 95.
Gadow 95.
Gage 82.
Gaupp 01, 06.
Gegenbaur 62.
Göppert 95, 96.
Hay 89, 90.
Hepburn 07.
Van der Hoeven 66.
Houghton 88.
Howes and Davies 88.
Lucas 86.
Meyerhofer 09.
Mivart 70.
Murray 97.
Nussbaum 08.
Parker 68, 77, 96.
Peter 94, 98.
Platt 97.
Reynolds 97.

Ryder 79.
Schauinsland 98.
Siegfanger 01.
Smellie 04 07.
Smith, F. 03.
Terry 04.
Van Pee 03, 03a, 04.
Waltou 67.
Whipple 96.
Wiedersheim 77, 92.
Wilder, H. H. 03.
Winslow 98.
Zwick 07.

4. Muscles.

Brons 07.
Davison 94.
Dräner 03.
Dugès 34.
Kyoleshymer 02a, 04.
Fürbringer 86.
Gage, S. P. 90.
Göppert 94.
Humphry 72.
Maeallum 87a.
Maurer 06.
Mivart 69, 69a.
Platt 97.
Ribbing 07.
Schaper 02.

5. Blood and Vascular System.

Boas 81.
Claypole 93, 96.
Crisp 60.
Essex 90.
Favaro 06.
Gage, S. H. 85a.
Gulliver 73.
Harting 58.
Hochstetter 06.
Van der Hoeven 41, 67a.
Joseph 05.
Kneeland 68.
Knoll 96.
Königstein 09.
Kossmann 03.
Langerhans 73.
Maeallum 02.
Mandl 39, 39a.
Marcinowski 07.
Miller 00a.
Minot 98, 00.
Owen 35, 42.
Roméinx 06.
Rosy 07.
Schmidt, H. D. 78.
Sobotta 02.

Vaillant 62.
Welikij 98.
Wyman 54.
Zuckerkandl 95, 95a.

6. Respiratory System.

Boerger 96.
Bridge 00.
Camerano 85a.
Clemens 95.
Druner bl.
Eycleshymer 95a.
Göppert 98.
Kneeland 54.
Königstein 95a.
Maurer 88, 88a.
Miller 90b, 92, 96.
Wiedersheim 77b.
Wilder, H. H. 96.
Wyman 54.

7. Digestive System.

Becker 92.
Beasley 60.
Blyer 71.
Bolau 99.
Broman 04.
Choronshitzky 97, 00.
Gage-Gage 90.
Gage, S. H. 85.
Göppert 91, 00.
Green 99, 97.
Holmgren 04.
Kingsbury 94.
Leydig 98.
Macallum 87.
Mc Gill 108a.
Mathews 99.
Meckel 19.
Miller 96.
Monti 98.
Nemiloff 02.
Oppel 96.
Platt 96.
Prenant 04.
Reese 04b.
Richter 92.
Riddle 09.
Saint-Hilaire 03.
Schulz 90.

8. Nervous System.

Allis 03.
Anderson 93.
Bender 95a.
Brachet 07.
Bugnion 74.
Carlson 06.

Deen 54.
Edinger 04.
Fischer, J. G. 43.
Galeotti 97.
Herrick 93, 94, 99.
Hirsch-Tabor 00.
Hoffmann 02.
Humphry 72.
Jelkile 97.
Johnston, J. B. 09, 06.
Kingsbury 95a.
Klaussner 83.
Kölsn 86.
Kupfer 06.
Levi 97.
Leydig 70a.
Macallum 87.
Mc Gill 08, 08a.
McGregor 96.
Miller 00.
Monti 99.
Nakagawa 91.
Nemiloff 02.
Neumayer 96.
Norris 00, 08, 09.
Osborn 83, 84, 86, 86a, 88.
Platt 96a.
Rabaschkin 01, 03.
Rage 97.
Sairnow 89.
Strong 92, 95.
Tuerckheim 03.
Vaillant 62.
Warren 05.

9. Sense Organs.

Bawden 94.
Beer 99.
Bugnion 73.
Dubois 90.
Eigenmann 00, 00b.
Eigenmann and Denny 00, 00a.
Eycleshymer 93.
Flemming 00.
Hess 89.
Howard 08.
Kingsbury 95, 95b, 95, 00.
Kingsley 00.
Kingsley and Rohlick 90.
Kohl 89, 91.
Kuhn 79.
Lauber 02.
Leydig 68, 76b.
Lveb 06.
Mallrann 75, 76.
Mall 93.
Mihalkovics 98.

Norris 94.
Okajima 95.
Okasina 96.
Rabl 98.
Reese 05a, 94a.
Rejsek 97a.
Renius 83, 94, 05.
Schlamp 94, 92.
Schulze, F. E. 93, 70.
Seydel 95.
Studnicka 97.
Takabashi 09.
Terry 06.
Wilder, H. H. 92.
Wilder, J. W. 08.

10. Integument.

Ancel 04.
Chauvin 84.
Eycleshymer 04b.
Grote 76.
Hoffenhain 93.
Leydig 73, 76, 76a.
Merkel 80.
Nidoga 93.
Rabl 95.
Reese 04b, 05.
Schoberg 93.
Studer 74.

11. Genito-urinary Organs.

Bidder 46.
Broman 04.
Drzewina 06.
Duvernoy 47.
Felix and Bühler 06.
Field 94.
Fürbringer 77.
Hall 04.
Heidenhain 90.
Johnston, C. 76.
Joseph 05a.
Kampen 08.
Kingsbury 97a.
Knappe 86.
McGregor 99.
Meyer 75.
Mintl 98.
Persol 89.
Ruthke 25.
Reese 04, 04a.
Valentin 41.

12. Variations.

Bompas 96.
Howes 93.
Waite 97.
Window 04.

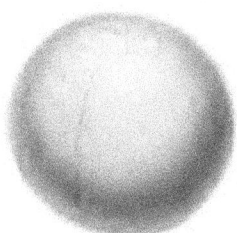

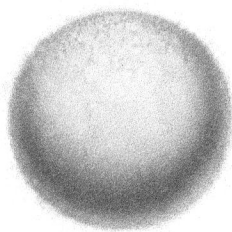

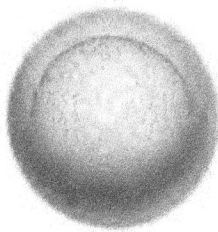

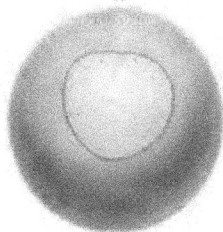

13 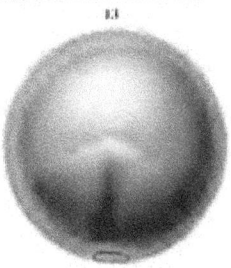 14 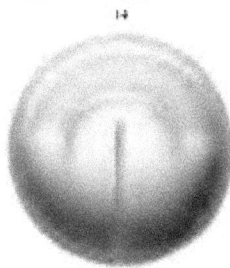 15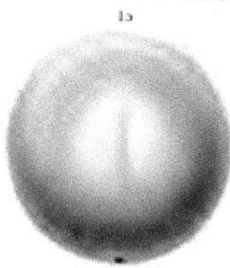

16  17  18

19 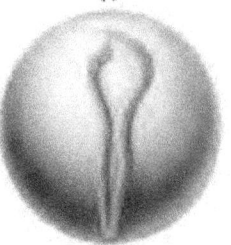 20  21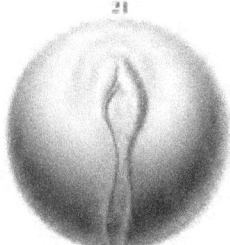

22  23  24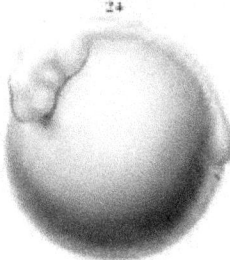

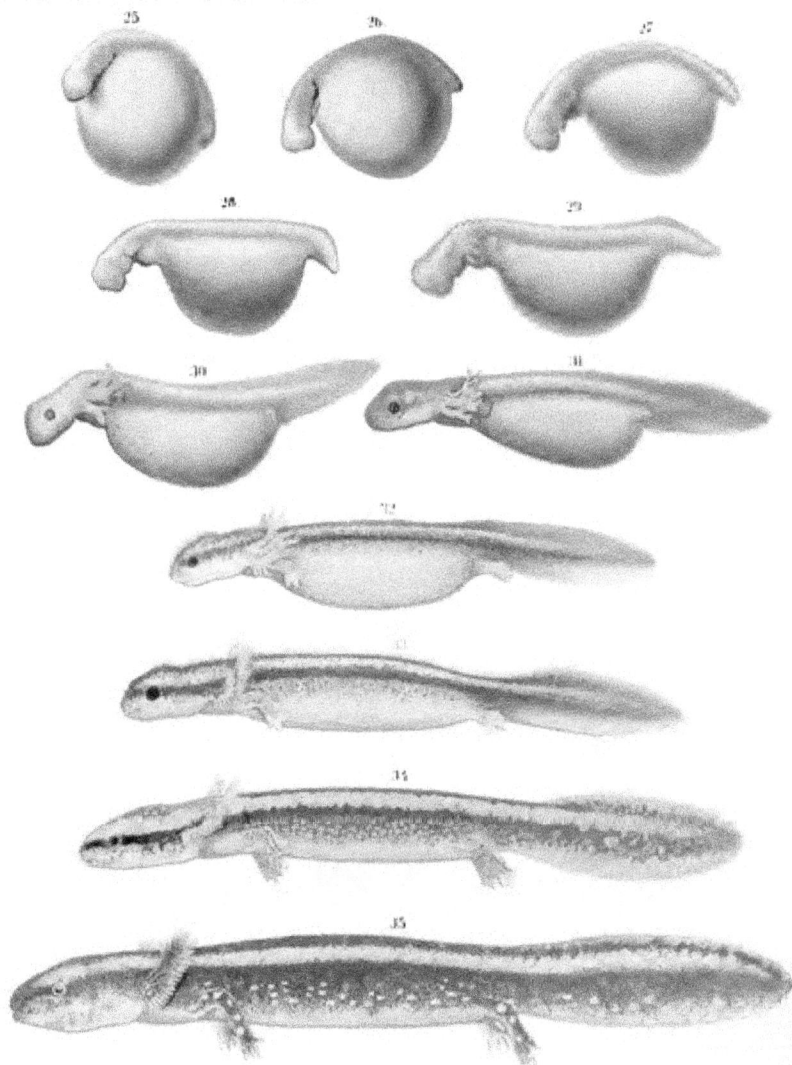